Elements of
Differentiable Dynamics and
Bifurcation Theory

Elements of Differentiable Dynamics and Bifurcation Theory

David Ruelle

Institut des Hautes Etudes Scientifiques
Bures-sur-Yvette, France

ACADEMIC PRESS, INC.
Harcourt Brace Jovanovich, Publishers

Boston   San Diego   New York
Berkeley   London   Sydney
Tokyo   Toronto

ACADEMIC PRESS, INC.
1250 Sixth Avenue, San Diego, CA 92101

United Kingdom Edition published by
ACADEMIC PRESS INC. (LONDON) LTD.
24–28 Oval Road, London NW1 7DX

Library of Congress Cataloging-in-Publication Data

Ruelle, David.
    Elements of differentiable dynamics and bifurcation theory / David
Ruelle.

        p.   cm.
    Bibliography: p.
    Includes index.
    ISBN 0-12-601710-7
    1. Differentiable dynamical systems.   2. Bifurcation theory.
I. Title.
QA614.8.R84 1988
515.3'5–dc 19   88-18097
                                                            CIP

89 90 91 92   9 8 7 6 5 4 3 2
Printed in the United States of America

Contents

Preface

The study of differentiable dynamical systems—differentiable dynamics—has been one of the most successful fields of mathematical research in the last twenty years. Many profound results have been uncovered and applied to other branches of mathematics as well as to the natural sciences. The explosive development of the subject has unfortunately led to a fragmentation into different subfields, to which it is becoming more and more difficult to gain access. This is particularly regrettable with regard to the applications and explains in part the low quality of much of the recent physics literature on "chaos." A solution to this problem, i.e., a unified presentation of all the important and useful results on differentiable dynamical systems, would at this time require a large treatise that could probably not be written by any single mathematician.

My ambition has been more modest: This monograph is intended only as an *introduction* to differentiable dynamics, with emphasis on bifurcation theory and hyperbolicity, as needed for the understanding of complicated time evolutions occurring in nature (turbulence and "chaos"). My aim is thus to present the basic facts of differentiable dynamics in an accessible manner for use by mathematicians or mathematically inclined students of the natural sciences. In order to get to the heart of the matter quickly, I emphasize ideas rather than proofs. Results are formulated precisely, and some of the proofs are given. Other proofs, especially the longer ones, are outlined only, or omitted, but references are provided. Since I have in view the applications to natural phenomena, my approach is, to some extent, unconventional, with more emphasis than usual on infinite dimensional systems, noninvertible maps, attractors, and bifurcation theory.

After going through the material presented here, the serious reader should be in a good position to proceed (if he or she wishes) to the study of more advanced mathematical topics. Such topics include the detailed theory of Axiom A diffeomorphisms and flows, the ergodic theory of differentiable dynamical systems, and the analysis of "difficult" bifurcations like the Feigenbaum bifurcation. The serious reader referred to above should also be better equipped to enter the treacherous jungle of the literature on chaos.

This monograph contains three parts. The first part is centered on differentiable dynamics and begins with the definitions of manifolds and differentiable dynamical systems. Then fixed points, periodic orbits, and their invariant manifolds are introduced. Finally, we discuss attractors, bifurcations, and generic properties.

The second part is centered on bifurcations. The elementary bifurcations (saddle node, flip, Hopf) for a fixed point, or for a periodic orbit of a map, are studied in detail, as well as related bifurcations for semiflows. This requires some results on normally hyperbolic invariant manifolds. From there we proceed to a general study of hyperbolic invariant sets with applications to homoclinic intersections, as well as some of the less elementary bifurcations.

The third part is a collection of appendices. Appendices A, B, and C merely collect definitions and results from various parts of analysis for easy reference in the main text. Appendix D is of a different nature: It presents the basic theory of Axiom A systems in the general setup of maps and semiflows while weakening the usual requirement of a compact manifold. This is somewhat specialized material, and is therefore relegated to an appendix.

A number of questions not treated in the main text are presented as problems at the end of Parts 1 and 2. It is recommended that the reader have a look at these problems, even if he or she does not wish to spend much time working them out in detail.

This monograph owes much to discussions with a number of colleagues, particularly Mike Shub and Jacob Palis. The latter read the manuscript carefully and made a number of detailed suggestions which I have largely followed. The manuscript, written in part at the I.H.E.S. in Bures-sur-Yvette, in part at Rutgers University, was completed while the author was visiting Cal Tech as a Fairchild scholar.

1 Differentiable Dynamical Systems

Nouum opus facere me cogis ex ueteri, ...
pius labor, sed periculosa praesumptio.

—*Hieronymus*

We begin the theory of differentiable dynamics with some definitions and results on the space, or *manifold*, on which the system acts, and the different types of dynamical systems considered. The essential concepts are presented here, and we refer to the appendices for more details. Fixed points, periodic orbits, and their invariant manifolds are then studied. (Their persistence will be analyzed in Part 2.) This part ends with a discussion of attractors, bifurcations, and generic properties.

1. Manifolds

Physics gives many examples of differentiable dynamical systems in infinite-dimensional Banach spaces. In fact, Banach spaces (finite- or infinite-dimensional) provide the natural framework for the study of differentiable maps.[1] Definitions and results concerning Banach spaces are collected in Appendix A.4, and A.5. The prime examples to keep in mind are $\mathbf{R}^m$ with the Euclidean norm, and real Hilbert space.

If E, F are Banach spaces, the *derivative* of $f : E \mapsto F$ is a function $f' : E \mapsto \mathcal{L}(E, F)$ (where $\mathcal{L}(E, F)$ is the space of linear operators $E \mapsto F$) defined as usual such that

$$\lim_{\|\xi\| \to 0, \, \xi \neq 0} \frac{\|f(x + \xi) - f(x) - f'(x)\xi\|}{\|\xi\|} = 0.$$

We also write $f'(x) = D_x f$. The function f is *of class* C^r if it has continuous derivatives up to order r. A C^0 function is thus simply a continuous function. One also defines C^∞ functions (infinitely differentiable), C^ω functions (real analytic), and $C^{(r,\alpha)}$ functions where the rth derivative is Hölder-continuous of exponent α (see Appendix B.1). One often uses *differentiable* or *smooth* in a vague sense, with the meaning of $C^\mathbf{r}$ for some $\mathbf{r} \geqslant 1$, i.e., $\mathbf{r} = r$ integer $\geqslant 1$, $\mathbf{r} = (r, \alpha)$ with r integer $\geqslant 1$ and $0 < \alpha \leqslant 1$, $r = \infty$, or $\mathbf{r} = \omega$. We write $\mathbf{r} - 1$ for $r - 1$, $(r - 1, \alpha)$, ∞, or ω, respectively. *Sufficiently smooth* means $C^\mathbf{r}$ for $\mathbf{r}$ sufficiently large. If $E = \mathbf{R}^m$, $F = \mathbf{R}^n$, a map $f : E \mapsto F$ is a family $(f_j)_{1 \leqslant j \leqslant n}$ of functions $\mathbf{R}^m \mapsto \mathbf{R}$. The rth derivative $f^{(r)}$ can be identified with the $n \times m^r$ array of the partial derivatives $\partial^r f_j / \partial x_{i_1} \cdots \partial x_{i_r}$. In particular, $D_x f$ is the $n \times m$ matrix $(\partial f_j / \partial x_i)$.

A $C^\mathbf{r}$ *manifold* is a space M obtained by gluing together pieces (open sets) of a Banach space E so that it makes sense to speak of $C^\mathbf{r}$ functions defined on M, or with values in M (see Appendix B.4).[2] For instance, an open subset of E is a smooth manifold. The infinite-dimensional manifolds which we shall use will mainly be open subsets of Banach spaces.

Gluing together pieces of $\mathbf{R}^m$ yields an m-dimensional manifold. Suppose, for instance, that M is a subset of $\mathbf{R}^n$ such that, near each $x \in M$, $n - m$ of the coordinates of a point of M can be expressed as $C^\mathbf{r}$ functions of the remaining m coordinates. Then M can be considered as an m-dimensional $C^\mathbf{r}$ manifold (a $C^\mathbf{r}$ submanifold of $\mathbf{R}^n$). Conversely, if M is a separable[3] m-dimensional $C^\mathbf{r}$ manifold, *Whitney's theorem*

---

[1] We use interchangeably the phrases *differentiable map* and *differentiable function*.

[2] Several different spaces E_α could be used instead of just one space E.

asserts that M can be realized as a C^∞ or even C^ω closed submanifold of $\mathbf{R}^{2m+1}$ (see Appendix B.4.5). One says that M is embedded as a submanifold of $\mathbf{R}^{2m+1}$. This gives a good intuitive picture of finite-dimensional smooth manifolds and why they are always assumed to be C^∞ or C^ω.

A compact manifold[4] is necessarily of finite dimension, and therefore realizable as a compact submanifold of Euclidean space. (A compact submanifold of $\mathbf{R}^n$ is bounded and closed in $\mathbf{R}^n$). Examples of compact manifolds are the *m-sphere* S^m and the *m-torus* T^m. The m-sphere is the submanifold of $\mathbf{R}^{m+1}$ defined by the equation

$$S^m = \left\{ (x_1, \ldots, x_{m+1}) \in \mathbf{R}^{m+1} : \sum_1^{m+1} x_i^2 = 1 \right\}.$$

The m-torus is obtained by identifying (i.e., gluing together) opposite faces of the m-cube

$$\{(x_1, \ldots, x_m) : 0 \leqslant x_i \leqslant 1 \qquad \text{for } i = 1, \ldots, m\}.$$

Notice that T^1 is the same thing as S^1 and that T^m is the product of m copies of T^1. The 2-torus T^2 has a familiar realization as submanifold of $\mathbf{R}^3$ (the surface of the annulus of Fig. 1). In many respects, compact manifolds have a simpler behavior than non-compact manifolds. Therefore, one tends to formulate theorems for the case of compact manifolds. When confronted with a problem of local nature on a finite-dimensional manifold, one may modify the manifold to make it compact. For instance, $\mathbf{R}^m$ is noncompact, but if we are interested in the unit ball $\{\|x\| \leqslant 1\}$, we can identify it with the southern hemisphere of S^m by the stereographic map $x \mapsto y$ (Fig. 2).

The manifolds which we have introduced are sometimes called manifolds without boundary. The annulus of Fig. 1 is an example of a manifold with boundary (the boundary is T^2).

---

[3] M is separable if there is a sequence (x_n) dense in M (see Appendix A.2). This technical condition is usually satisfied in practice. It holds, for instance, when M is compact.

[4] A compact manifold is a manifold that is compact as a topological space (see Appendix A.2).

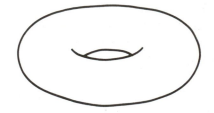

FIG. 1.   The 2-torus T^2 as boundary of an annulus (solid torus) in $\mathbf{R}^3$.

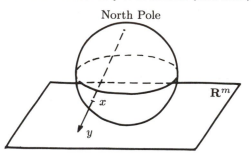

FIG. 2.   Stereographic projection of the sphere S^2 (minus one point) to the plane $\mathbf{R}^2$.

2. Differentiable Dynamics

The time evolution of natural systems is often given by a differential equation of the general form

$$(2.1) \qquad \frac{dx}{dt} = X(x).$$

In hydrodynamics, for instance, x would belong to a suitable functional space, and X would be a nonlinear partial differential operator (supplemented by boundary conditions). In other examples, like classical mechanics, x has to be interpreted as a vector in finite dimension, say in $\mathbf{R}^n$. More precisely, x belongs to a submanifold of $\mathbf{R}^n$, corresponding to conservation laws and, possibly, constraints.

In general, we shall assume that time evolution is given by a family $(f^t)_{t \geqslant 0}$ of maps of a differentiable manifold M, so that $f^t x$ describes the system at time t for initial condition x (we thus have $f^0 x = x$, i.e., f^0 is the identity map). We have supposed that t does not explicitly occur in the right-hand side of (2.1). This corresponds to assuming the *semigroup property* $f^{s+t} = f^s \circ f^t$ for the family (f^t).

We then say that $(f^t)_{t \geq 0}$ is a *semiflow* on M; it defines a *continuous time dynamical system*. Let us restrict the time to integer values, in terms of some arbitrary unit (microsecond, year, or million years), then the time evolution is described by the iterates of a single map $f = f^1$. Such a *discrete time dynamical system* also occurs if we study at integer values of t a system like (2.1) where the right-hand side $X(x,t)$ now contains t explicitly, but with period unity $X(x, t + 1) = X(x,t)$. An example of this is provided by the dynamics of populations of n interacting species subjected to seasonal influences of a one-year period.

The maps f^t are often also defined for negative t and form a group rather than just a semigroup. In particular, f^{-t} is the inverse of f^t. For continuous time, we then have a *flow* $(f^t)_{t \in \mathbf{R}}$ on M, while a discrete time dynamical system is defined by an invertible map f.

One would expect to obtain interesting statements on the time evolution defined by (2.1) only if very specific assumptions are made on the right-hand side X. This expectation turns out to be wrong: Remarkable phenomena (like the bifurcations discussed in Part 2) occur and can be analyzed in dynamical systems of great generality. The essential assumption that we shall make is that (f^t) is a *differentiable* dynamical system. This means that each $f^t : M \mapsto M$ is smooth of class C^r. (For the continuous-time case, mild conditions on the t-dependence will be added.) The differentiability condition will often be complemented by genericity assumptions (see Section 8.7). This means that some special (degenerate) situations are disregarded.

It is appropriate at this point to question the smoothness assumption for dynamical systems describing natural phenomena. There is a widespread view that functions that are significant for physics, or other natural sciences, are continuous, differentiable, and, in fact, analytic (apart from possible isolated singularities with definite meaning). This can be a somewhat dangerous philosophy because the study of very smooth dynamical systems leads naturally to very irregular functions.[5] It is true that standard phenomenological laws that are assumed in time evolution equations normally yield smooth dynamical systems. One should, however, be alert to nonsmooth behavior: In this respect, the discussion of hydrodynamical time evolution leads to questions that have not yet been resolved.

---

[5] See, for instance, the study of the rotation number in Section 13.2.

We now list some classes of differentiable dynamical systems that will recur in later discussion.

2.1. C^r Maps

A discrete time dynamical system, defined only for positive times, is determined by the *smooth map* $f = f^1$, which is, in general, not invertible (or does not have a smooth inverse).

2.2. C^r Diffeomorphisms

A discrete time dynamical system, defined for positive and negative times, is determined by a *diffeomorphism*, i.e., a smooth map f with smooth inverse. If a C^r map f has a C^1 inverse f^{-1}, then f^{-1} is, in fact, C^r, and f is thus a C^r diffeomorphism. This follows from the inverse function theorem (see Corollary B.3.2).

2.3. C^r Semiflows

A continuous time dynamical system, defined only for positive times, is a semiflow, i.e., a family $(f^t)_{t \geqslant 0}$ of C^r maps satisfying $f^0 = $ identity and $f^{s+t} = f^s \circ f^t$ (semigroup property). Various assumptions can be made on the t-dependance of f^t (at least some measurability, or continuity, or smoothness).

2.4. C^r Flows

A continuous time dynamical system, defined for positive and negative times, is a flow, i.e., a family $(f^t)_{t \in \mathbf{R}}$ of C^r diffeomorphisms satisfying $f^0 = $ identity and $f^{s+t} = f^s \circ f^t$ (group property). A flow may be considered as a semiflow by ignoring all f^t with negative t. The full flow can be reconstructed from the semiflow by using $f^{-t} = (f^t)^{-1}$. A semiflow $(f^t)_{t \geqslant 0}$ fails to be a flow if some f^t is not a diffeomorphism (because f^t is not surjective, not injective, or else because its inverse is not smooth).

2.5. Systems with a Group of Symmetries

Let G be a group of transformations of M; G is a *group of symmetries* for the dynamical system (f^t) if $f^t \circ g = g \circ f^t$ for all $g \in G$ and t. (One also says that (f^t) is *G-equivariant*). The existence of a symmetry group has to be carefully noted because it changes the expected (or *generic*) properties of dynamical systems. For instance, let $M = \mathbf{R}$ and let G be generated by the reflection $x \mapsto -x$, then a G-equivariant map f necessarily leaves the origin fixed.

2.6. Hamiltonian Systems

A frictionless mechanical system that is not submitted to external forces has a time evolution described by Hamilton's equations

$$\frac{dq_i}{dt} = \frac{\partial H}{\partial p_i}, \quad \frac{dp_i}{dt} = -\frac{\partial H}{\partial q_i} \qquad \text{for } i = 1, \ldots, n$$

in the $2n$-dimensional space (phase space) parametrized by the "position" coordinates $q_1, \ldots, q_n$ and the "momentum" coordinates $p_1, \ldots, p_n$. The *Hamiltonian function* H depends smoothly on these $2n$ coordinates (but not explicitly on time). The integer n is the *number of degrees of freedom* of the Hamiltonian system. Since H is a constant of the motion (conservation of energy), the phase space is decomposed into a family of subsets $H = $ constant (*energy surfaces*), which are invariant under time evolution. The volume element $dq_1 \cdots dq_n \, dp_1 \cdots dp_n$ is also preserved by time evolution (*Liouville's theorem*). This shows that Hamiltonian systems, also called *conservative systems*, are very special dynamical systems. By contrast, systems with friction, or *dissipative systems*, have a more "normal" or generic behavior.

2.7. Topologies for Dynamical Systems on a Compact Manifold

Let $\mathcal{D}$ be a space of differentiable dynamical systems on a manifold M, as discussed above. (For the moment, M may be an infinite-dimensional Banach manifold.) To define a *topology* on $\mathcal{D}$, we have to say what the *neighborhoods* of every element of $\mathcal{D}$ are (see Appendix A.2). We shall discuss the example of the space $\mathcal{D}$ of $C^{\mathbf{r}}$ maps and define the neighborhoods of a $C^{\mathbf{r}}$ map $f_0 : M \mapsto M$ in $\mathcal{D}$. We may assume that M is obtained by gluing together pieces U_i, or pieces U_i', of Banach spaces, and that f_0 sends U_i inside U_i'. To be specific, let the U_i, U_i' be open and the closure of $f_0 U_i$ be contained in U_i'. We have thus replaced the map $f_0 : M \mapsto M$ by a collection of Banach space maps $f_0|U_i$, for which topologies are defined (see Appendix B.2).[6] If the $\mathcal{N}_i$ are $C^{\mathbf{r}}$

---

[6] More precisely, we can choose the U_i such that each $f_0|U_i$ is in the space $C^{\mathbf{r}}(U_i, E_i')$, where E_i' is the Banach space containing U_i' as an open set.

neighborhoods of the $f_0|U_i$, we define

$$\mathcal{N} = \{f \in \mathcal{D} : fU_i \subset U_i' \quad \text{and} \quad f|U_i \in \mathcal{N}_i \qquad \text{for all } i\}.$$

We may now declare that the neighborhoods of f_0 are all the subsets of $\mathcal{D}$ that contain an $\mathcal{N}$ as above when the family (U_i) is allowed to vary over a suitable class of open coverings of M.

Naturally, different choices of a "suitable class" of open coverings (U_i) may lead to different topologies. Because of this complication, we shall in general not put any topology on the space of dynamical systems on a manifold M, except when M is compact.[7] When M is compact, every finite open cover (U_i) of M yields the same topology, as one can easily check, and we call this the $C^{\mathbf{r}}$ *topology*. [For integer $\mathbf{r}$, this is the topology of uniform convergence of the maps $M \mapsto M$ and of their derivatives up to order $\mathbf{r}$. For other $\mathbf{r}$'s there are the usual complications.] We denote by $C^{\mathbf{r}}(M, M)$ the space of $C^{\mathbf{r}}$ maps $M \mapsto M$ with the $C^{\mathbf{r}}$ topology; the latter can also be defined by a metric, as we shall see in Section 8.8.

We call $\text{Diff}^{\mathbf{r}}(M)$ the space of $C^{\mathbf{r}}$ diffeomorphisms of the compact manifold M with the $C^{\mathbf{r}}$ topology, i.e., the topology induced by the topology of $C^{\mathbf{r}}(M, M)$. [$\text{Diff}^{\mathbf{r}}(M)$ is an open subset of $C^{\mathbf{r}}(M, M)$, as one can see by using the inverse function theorem B.3.1.]

Note that composing maps: $(f, g) \mapsto f \circ g$, or inversion: $f \mapsto f^{-1}$, are continuous operations for the $C^{\mathbf{r}}$ topologies (the latter when restricted to $\text{Diff}^{\mathbf{r}}(M)$).

For continuous time dynamical systems, various topologies can be defined by considering the maps $M \times [0, 1] \mapsto M$ (semiflows) or $M \times [-1, 1] \mapsto M$ (flows). In particular, if M is compact, we define $\mathcal{F}^{\mathbf{r}}(M)$ to be the space of flows on M such that $(x, t) \mapsto f^t x$ is of class $C^{\mathbf{r}}$. For integer $\mathbf{r} \geqslant 1$, the topology on $\mathcal{F}^{\mathbf{r}}(M)$ is taken to be the topology of uniform convergence on $M \times [-1, 1]$—or, equivalently, on compacts of $M \times \mathbf{R}$—of the map $(x, t) \mapsto f^t x$ and of its derivatives (with respect to x, t) up to order $\mathbf{r}$. The extension to the case where $\mathbf{r}$ is not an integer is obvious.

Flows on a compact manifold are often obtained by integrating vector fields (see next section): the space $\mathcal{X}^{\mathbf{r}}(M)$ of $C^{\mathbf{r}}$ vector fields has a very

---

[7] Note, however, that the spaces $C^{\mathbf{r}}(U, F)$ of Appendix B.1 are natural for local studies, and we shall use them to define "local" topologies in a neighborhood of a point or of a compact set.

natural topology because it is locally a map $\mathbf{R}^m \mapsto \mathbf{R}^m$. We shall see in Section 8.8 how to define this topology by a metric.

3. Vector Fields

We now return to the differential equation

$$(3.1) \qquad \frac{dx}{dt} = X(x)$$

for a more detailed discussion. We first consider the case where Eq. (1.1) is defined in an open subset U of a Banach space E, and where $X : U \mapsto E$ is a smooth function, i.e., X is of class $C^{\mathbf{s}}$, $\mathbf{s} \geqslant 1$.[8] We then have *local existence and uniqueness* of solutions of (1.1): Given $a \in U$, there are $\varepsilon, T > 0$ such that (1.1) has a unique solution $t \mapsto f^t x$ in the interval $(-T, T)$ for an initial condition $x \in E_a(\varepsilon)$ (ball of radius ε centered at a). Furthermore, $f^t x$ is a smooth function of x and t (in fact, a $C^{\mathbf{s}}$ function, see Appendix B.6).

It is a fact that the evolution equations encountered in physics often do not correspond to a smooth $X : U \mapsto E$, except if E has finite dimension. Consider for instance the *diffusion equation*

$$(3.2) \qquad \frac{\partial x}{\partial t} = \kappa \Delta x,$$

where $\xi \mapsto x(\xi)$ is a real function on a subset of $\mathbf{R}^3$, $\kappa > 0$ is a *diffusion coefficient*, and $\Delta = \sum_{i=1}^{3} \frac{\partial^2}{\partial \xi_i^2}$ is the Laplace operator. (If x is a temperature field, this is the *heat equation*.) If one interprets (3.2) as an evolution equation in some Banach space of differentiable functions, one has to face the fact that the Laplace operator lowers differentiability. Any way one looks at it, the function $x \mapsto \Delta x$ is not smooth. The above local existence and uniqueness theorem does not apply, and one has to resort to other tools to obtain a dynamical system from (3.2) (actually, one obtains a semiflow, not a flow). These comments on the diffusion equation also apply to the equations of the hydrodynamics of viscous fluids, as well as equations for other dissipative systems.

In practice, the local existence and uniqueness theorem for solutions of (3.1) is mostly of interest in finite dimensions. There, however, one

---

[8] Actually, $\mathbf{s} \geqslant (0,1)$ is what is required throughout this section.

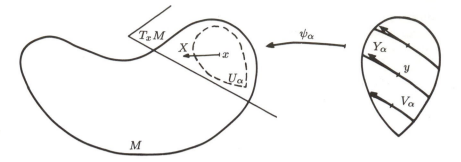

FIG. 3.   Vector field X on a manifold M.

would like to treat the case of manifolds, rather than just of open subsets of $\mathbf{R}^m$. An m-dimensional manifold M can be viewed as a submanifold of $\mathbf{R}^n$ (one can take $n = 2m + 1$ by Whitney's Theorem). For each $x \in M$, there is a *tangent space* $T_x M$ that is an m-dimensional linear space. (We view it here as a subspace of $\mathbf{R}^n$.) A *vector field* X on M is a function associating with each $x \in M$ a *tangent vector* $X(x)$ at x; thus $X(x) \in T_x M$ for all x. Remember that an m-dimensional manifold is obtained by gluing together pieces of $\mathbf{R}^m$. There are thus open sets $U_\alpha \subset M$, open sets $V_\alpha \subset \mathbf{R}^m$, and maps $\psi_\alpha : V_\alpha \mapsto \mathbf{R}^n$ such that ψ_α identifies V_α with the set U_α (considered as subset of the ambient space $\mathbf{R}^n$). For a vector field X on M, there will then be functions $Y_\alpha : V_\alpha \mapsto \mathbf{R}^m$ such that

$$X(\psi_\alpha y) = (D_y \psi_\alpha) Y_\alpha(y),$$

where $D_y \psi_\alpha$ is the $n \times m$ matrix of partial derivatives of the n components of ψ_α by the m components of y. We have thus replaced a vector field X on M by "flat" vector fields Y_α on the V_α. Notice that

(3.3) $$Y_\beta(\psi_\beta^{-1}\psi_\alpha y) = (D_y(\psi_\beta^{-1}\psi_\alpha)) Y_\alpha(y)$$

for $\psi_\alpha y \in U_\alpha \cap U_\beta$. This is the familiar formula for the transformation of a vector field in a change of coordinates $y \mapsto \psi_\beta^{-1}\psi_\alpha y$, where $D_y(\psi_\beta^{-1}\psi_\alpha)$ is the $m \times m$ Jacobian matrix corresponding to the change of coordinates. The vector field X is of class $C^\mathbf{s}$ if the Y_α are $C^\mathbf{s}$ functions and the manifold M is at least of class $C^{\mathbf{s}+1}$.[9]

---

[9] For a $C^\mathbf{r}$ manifold, the functions $\psi_\beta^{-1}\psi_\alpha$ are of class $C^\mathbf{r}$, and $y \mapsto D_y(\psi_\beta^{-1}\psi_\alpha)$ is of class $C^{\mathbf{r}-1}$. Formula (3.3) requires that $\mathbf{s} \leqslant \mathbf{r} - 1$ in order to make sense. We shall usually take M of class C^∞ or C^ω, which automatically satisfies the above condition.

One can use the change of coordinates formula to define tangent vectors and vector fields "abstractly," i.e., without reference to an embedding of the manifold M in $\mathbf{R}^n$, and one can then also treat Banach manifolds (see Appendix B.5, B.6). If one uses the convenient picture of a manifold embedded in $\mathbf{R}^n$ (see Fig. 3), one should be careful to consider tangent spaces $T_x M$, $T_{x'} M$ at different points as *disjoint*, although they may intersect when realized as subspaces of $\mathbf{R}^n$. ($T_x M$ and $T_{x'} M$ are sets of vectors with different origins.) The union TM of the $T_x M$ is called the *tangent bundle*. (TM may be viewed as a $2m$-dimensional manifold, of class C^{r-1} if M is of class C^r.)

If X is a smooth vector field on M, a solution of the differential equation

$$(3.4) \qquad \frac{dx}{dt} = X(x) \quad \text{on } M$$

is a function defined on an open interval (T_-, T_+) of $\mathbf{R}$, with values in M such that, on subintervals covering (T_-, T_+), it is of the form

$$t \mapsto \psi_\alpha(y_\alpha(t)),$$

where y_α satisfies the equation $dy_\alpha/dt = Y_\alpha(y_\alpha)$ in V_α. The compatibility condition (3.3) ensures that this is a good definition. A solution of (3.4) is also called an *integral curve* of X. A differential equation on an m-dimensional manifold is thus locally the same thing as a differential equation in $\mathbf{R}^m$, and we therefore have again local existence and uniqueness.

For a manifold M consisting of an open set U in a Banach space E, one can take as $T_x M$ a copy of E for each $x \in U$. The tangent bundle is then identified with $U \times E$, and a C^r function $U \mapsto E$ corresponds to a C^r vector field.

Both in the finite- and infinite-dimensional cases, an integral curve of X is everywhere tangent to X, and $X(f^t x)$ is the velocity vector of the motion $t \mapsto f^t x$. By local existence and uniqueness, each initial condition $x \in M$ determines a solution $t \mapsto f^t x$ of (1.4) on a maximal interval $(T_-(x), T_+(x)) \ni 0$. This solution is unique. One shows that the set

$$\Gamma = \{(x,t) \in M \times \mathbf{R} : t \in (T_-(x), T_+(x))\}$$

is open in $M \times \mathbf{R}$, and the map $(x,t) \mapsto f^t x$ is C^s on Γ if X is C^s. The property $f^{s+t} = f^s \circ f^t$ holds where it makes sense, and (f^t) is called a *local flow*.

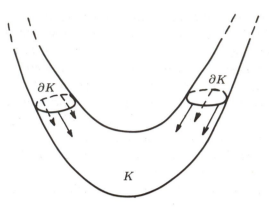

FIG. 4.   Vector field defining a semiflow on a compact subset K of a manifold.

If M is a compact manifold, the vector field X is bounded[10] and $f^t x$ cannot go to infinity. This implies that $f^t x$ is defined for all $t \in \mathbf{R}$. In other words, $\Gamma = M \times \mathbf{R}$ and (f^t) is a C^s *flow*.

As another example, assume that there is a compact subset K of M, bounded by a submanifold ∂K such that, at each point of ∂K, the vector field X points inside K. Then $f^t x$ is defined for all $x \in K$, $t \geqslant 0$: we have a semiflow on K (Figure 4).

We have not discussed the "general" case of a time-dependent differential equation depending on a parameter λ:

$$(3.5) \qquad \frac{dx}{dt} = X(x, t, \lambda),$$

where t belongs to an interval I and λ to a manifold Λ. The reader will check that the study of (3.5) reduces to that of the time-independent equation

$$\frac{d\tilde{x}}{dt} = \tilde{X}(\tilde{x})$$

on $M \times I \times \Lambda$, with $\tilde{X}(x, t, \lambda) = (X(x, t, \lambda), 1, 0)$.

Notice that one can define a space of smooth flows either by imposing differentiability conditions with respect to x and t on the family $(f^t x)$,

---

[10]The length of a tangent vector $X(x)$ may be defined by reference to the Euclidean metric of a space $\mathbf{R}^n$ in which M is embedded.

or by imposing differentiability conditions on the vector field X, which generates (f^t). In the case of a compact manifold, the space $\mathcal{F}^r(M)$ of C^r flows on M, and the space $\mathcal{X}^r(M)$ of C^r vector fields have been introduced in Section 2.7. A vector field in $\mathcal{X}^r$ defines a flow in $\mathcal{F}^r$, but a flow in $\mathcal{F}^r$ is associated with a vector field that is in general only in $\mathcal{X}^{r-1}$.

4. Fixed Points and Periodic Orbits. Poincaré Map

Let (f^t) be a dynamical system acting on a manifold M. (We always assume that $(x,t) \mapsto f^t x$ is continuous.) The *orbit* of a point x of M is the set $\{f^t x\}$, where t takes the values for which f^t is defined: reals or integers, possibly restricted to $t \geqslant 0$. The *forward orbit* is $\{f^t : t \geqslant 0\}$ and the *backward orbit* is $\{f^t x : t \leqslant 0\}$ (if defined).

One says that a is a *fixed point* if its orbit consists of a only. For a map f, this means $fa = a$. In the case of a flow or semiflow (f^t) obtained by integrating a vector field X, a is a fixed point if and only if $X(a) = 0$, i.e., if a is a *critical point* of X (we shall also say a *fixed point* of X).

One says that a is a *periodic point* if there is some $t > 0$ such that $f^t a = a$. The lower bound of such t is the *period* of a, and we denote it by $T(a) \geqslant 0$. When a is a fixed point, $T(a) = 1$ in the discrete time case, and $T(a) = 0$ in the continuous time case. The set $\{f^t a\} = \{f^t a : 0 \leqslant t \leqslant T(a)\}$ is called a *periodic orbit* (or *closed orbit*). For a discrete time dynamical system it is a finite set, for continuous time it is continuously infinite when $T(a) \neq 0$.

For a discrete time dynamical system, the periodic points for f are precisely the fixed points for some iterate f^n of f, with $n = 1, 2 \ldots$. This largely reduces the study of periodic points to the study of fixed points.

In the continuous-time case, a periodic point a with $T(a) > 0$ is a fixed point for $f^{T(a)}$, but $f^{T(a)}$ has a continuum of other fixed points near a, namely the other points on the orbit of a. Because of this unusual or *nongeneric* feature of the map $f^{T(a)}$, another map is often preferred as an aid in the study of the periodic orbit through a. This is the *Poincaré return map*, which we now define.

Let a be a periodic point of period $T(a) > 0$ for a semiflow (f^t). We assume that $(x,t) \mapsto f^t x$ is smooth, i.e., C^r with $r \geqslant 1$, in a neighborhood of $(a, T(a))$. We saw in Section 3 that this will be the case if (f^t) is obtained by integration of a smooth vector field.

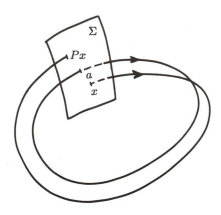

FIG. 5. Poincaré map.

Smoothness of $(x, t) \mapsto f^t x$ will also hold for the solutions of some important equations in Banach spaces that have a nonsmooth right-hand side.

Let Σ be a smooth codimension-1 submanifold containing a and transversal to the orbit of a (think of a hypersurface that is not tangent to the orbit; an exact definition will be given in a minute). Σ is a *local cross-section* of the flow. If x is close to a, one can choose t close to $T(a)$ such that $f^t x \in \Sigma$ (see Fig. 5). This defines a map $P : x \mapsto f^t x$, called the *Poincaré map*, or *first return map*, from a neighborhood of a in Σ to Σ. The map P has a as a fixed point and will turn out to be of class C^r, i.e., as differentiable as the original data.

We now use some definitions and results from Appendix B to make a more precise analysis. Let (U, φ, E) be a C^r *chart* of our manifold around x. This means that U is an open subset of M, $U \ni x$, and φ is a C^r diffeomorphism of U to an open subset of the Banach space E. The *tangent space* $T_x M$ is a linear space defined to be isomorphic to the Banach space E by some isomorphism $T_x \varphi : T_x M \mapsto (\varphi x, E)$.[11] Given another chart (U', φ', E') around x, the following change of coordinate formula must hold:

$$(T_x \varphi')(T_x \varphi)^{-1} = (\varphi' \varphi^{-1}, D_{\varphi x}(\varphi' \varphi^{-1})).$$

This agrees with what we said in Section 3 about finite-dimensional manifolds and open subsets of Banach spaces. The norms defined on

---

[11] In the pair $(\varphi x, E)$, φx is just an index, we take a different copy of E for each φx.

$T_x M$ by the various isomorphisms $T_x\varphi$ are equivalent. The union of the $T_x M$ forms a manifold TM called the *tangent bundle* (see Appendix B.5). Let $f : M \mapsto N$ be a C^r map of M to another C^r manifold N, and let (V, ψ, F) be a C^r chart of N around fx. The *tangent map* to f at x is the linear map $T_x f : T_x M \mapsto T_{fx} N$ defined by

$$(4.1) \qquad (T_{fx}\psi)(T_x f)(T_x\varphi)^{-1} = (\psi f \varphi^{-1}, D_{\varphi x}(\psi f \varphi^{-1})).$$

This map plays the role of the derivative of f at x. If $g : N \mapsto Q$ is another C^r map, we have the formula $T_x(g \circ f) = (T_{fx}g)(T_x f)$ corresponding to the *chain rule*. The map $Tf : TM \mapsto TN$ is naturally defined to have the restriction $T_x f$ to $T_x M$; it is a C^{r-1} map (see Appendix B.5).

A C^r *submanifold* of M is a subset Σ that looks linear locally in suitable coordinates. For each $x \in \Sigma$, there is thus a C^r chart $(U, \varphi, F \times G)$ of M around x, where F and G are Banach spaces, such that $\varphi(x) = (0,0)$, and

$$\varphi(\Sigma \cap U) = (F \times \{0\}) \cap \varphi(U).$$

The tangent space to Σ at x is a subspace of $T_x M$ defined by $T_x \Sigma = (T_x\varphi)^{-1}(F \times \{0\})$. The *codimension* of Σ is by definition the dimension of G. In particular, codim $\Sigma = 1$ corresponds to taking $G = \mathbf{R}$. Calling φ_2 the component of φ along the factor $G = \mathbf{R}$, we have

$$\Sigma \cap U = \{y \in U : \varphi_2(y) = 0\},$$

i.e., Σ is defined on U by the vanishing of φ_2. We also have

$$(4.2) \qquad\qquad T_x \Sigma = (T_x\varphi_2)^{-1}\{0\},$$

where $T_x\varphi_2 : T_x M \mapsto \mathbf{R}$ is a continuous linear functional on $T_x M$.

We are now in a position to discuss in more detail the construction of the Poincaré return map. The assumed smoothness of $(x, t) \mapsto f^t x$ implies the existence of $\frac{d}{dt}f^t a$ at $t = T(a)$, defining a vector X.[12] Furthermore, since a is not a fixed point, we have $X \neq 0$. One version of the implicit function theorem (Corollary B.3.4) says that the piece of orbit $Q = \{f^t a : |t - T(a)| < \varepsilon\}$ for a suitable $\varepsilon > 0$ is a C^r submanifold of M, of dimension 1, tangent at a to $\frac{d}{dt}f^t a|_{t=T(a)} = X \neq 0$.

---

[12] The velocity vector $\frac{d}{dt}f^t x$ can be formally defined by taking the tangent map to $t \mapsto f^t x$, applied to $(t,1) \in T_t\mathbf{R}$, where the tangent space to $\mathbf{R}$ is identified with $\mathbf{R} \times \mathbf{R}$.

In general, the submanifolds Σ and Q of M are defined to be *transversal* at $a \in \Sigma \cap Q$ (see Appendix B.4) if the tangent spaces satisfy $T_a\Sigma + T_aQ = T_aM$. [If Σ and Q both had infinite dimension, one would also impose that $T_a\Sigma \cap T_aQ$ had a closed complement in T_aM]. In our case, codim $\Sigma = 1$, dim $Q = 1$, and T_xQ is spanned by X. Transversality thus means that

$$X \notin T_a\Sigma.$$

This will be our assumption in studying the Poincaré map. Using (4.2), we may rewrite this condition as

$$(T_a\varphi_2)X \neq 0.$$

We now apply the implicit function theorem B.3.3 to the C^r map $(x, t) \mapsto \varphi_2(f^t x)$, defined in a neighborhood of $(a, T(a))$ in $M \times \mathbf{R}$ and with values in a neighborhood of 0 in $\mathbf{R}$. We have

$$\frac{d}{dt}\varphi_2(f^t a)|_{t=T(a)} = (T_a\varphi_2)\left(\frac{d}{dt}f^t a|_{t=T(a)}\right)$$
$$= (T_a\varphi_2)X \neq 0.$$

Therefore, there is a unique function τ from a neighborhood of a in M to a neighborhood of $T(a)$ in $\mathbf{R}$ such that $\varphi_2(f^{\tau(x)}x) = 0$, i.e., $f^{\tau(x)} \in \Sigma$. This function is C^r: If we restrict it to Σ, we have a C^r function that may be interpreted as *first return time*. The *first return map* $P : x \mapsto f^{\tau(x)}x$ defined in a neighborhood of x on Σ is thus also C^r, which is what we wanted to prove.

Note that the spirit of the above proof is the following: If a manifold or function has a well-defined linear approximation at a point, and if it is defined from C^r data, then this manifold or function can be shown to be C^r near the point by use of the implicit function theorem. Such use of the implicit function theorem (in one form or another) soon becomes automatic, and one says that the result is "obvious."

4.1. Proposition (on the Poincaré return map). *Let (f^t) be a semiflow on the C^r manifold M, and a a periodic point of period $T(a) > 0$. We assume that $(x, t) \mapsto f^t x$ is continuous, and that in a neighborhood of $(a, T(a))$, the map $(x, t) \mapsto f^t x$ is C^r, $\mathbf{r} \geq 1$. Let Σ be a C^r submanifold of codimension 1 transversal at a to the orbit of a. For x near a in M, there is a unique $\tau(x)$ near $T(a)$ in $\mathbf{R}$ such that $f^{\tau(x)}x \in \Sigma$. The*

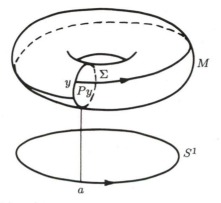

FIG. 6. Global Poincaré map.

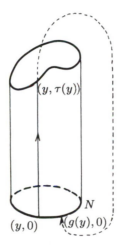

FIG. 7. Suspension of a map g with roof function τ.

function τ is $C^{\mathbf{r}}$, and the first return map $P : x \mapsto f^{\tau(x)}x$ is $C^{\mathbf{r}}$ from a neighborhood of a in Σ to Σ.

We have constructed a *local* Poincaré map for a semiflow having a periodic orbit. A Poincaré map, however, can be defined in other situations as well. For instance, let M be compact, and suppose that there is a map $\pi : M \mapsto S^1$ of M to a circle and a vector field X on M such that $(T\pi)X$ never vanishes (see Fig. 6). Given $a \in S^1$, a *global* first return map $P : \Sigma \mapsto \Sigma$ is defined by the global cross-section $\Sigma = \pi^{-1}a$ for the flow defined by X; P is a $C^{\mathbf{r}}$ diffeomorphism.

As a converse to the above construction, one may *suspend* a $C^\mathbf{r}$ diffeo-
morphism $g : N \mapsto N$ to obtain a $C^\mathbf{r}$ flow. Let $\tau : N \mapsto \mathbf{R}$ be a $C^\mathbf{r}$ function, positive, bounded, and bounded away from zero (roof function).
Consider the set $\{(y, \ell) \in N \times \mathbf{R} : 0 \leqslant \ell \leqslant \tau(x)\}$ and glue $(y, \tau(y))$ to
$(gy, 0)$; a $C^\mathbf{r}$ manifold M is obtained in this way (see Fig. 7). A flow
on M is defined by $((y, \ell), t) \mapsto f^t(y, \ell) = (y, \ell + t)$, when $0 \leqslant \ell + t \leqslant$
$\tau(y)$, and continued in an obvious way to other values of t, using the
identification $(y, \tau(y)) \equiv (gy, 0)$. This flow is $C^\mathbf{r}$ and corresponds to
the vector field $X = (0, 1)$.[13] The manifold M is called *suspension* of
N by g; the flow (f^t) is the *special flow* over g with *roof function* τ.
(One also says that (f^t) is the suspension of g, particularly if $\tau = 1$.)
By identifying N with the cross-section $\Sigma = (N, 0)$ of the flow (f^t), we
obtain g as a Poincaré map. The Poincaré map and the suspension are
thus inverse constructions. They do not give an exact correspondence
between flows and diffeomorphisms because a flow does not always have
a global cross-section. Nevertheless, the correspondence points to the
important fact that if some phenomenon occurs for diffeomorphisms in
dimension $\geqslant m$, the corresponding phenomenon for flows occurs only
in dimension $\geqslant m + 1$.

Given a semiflow (f^t), for example the flow corresponding to a vector field, one can always associate with it the *time one map* f^1. Note,
however, that a map is not in general the time one map of a semiflow.
[For instance, an orientation reversing diffeomorphism of S^1 cannot be
the time one map of a flow. See also Problem 7.]

5. Hyperbolic Fixed Points and Periodic Orbits

We shall successively discuss fixed points and periodic orbits for maps
and semiflows.

5.1. Fixed Points of a Map

We say that a fixed point a of a smooth map $f : M \mapsto M$ is *hyperbolic* if
the tangent map $T_a f : T_a M \mapsto T_a M$ is hyperbolic, i.e., if the spectrum
of $T_a f$ is disjoint from the unit circle $\{z : |z| = 1\}$. (See Appendix A.5
for a definition of the spectrum.) In particular, a is called *attracting* if

---

[13]Equivalently, one could use the roof function 1 and a vector field X such that
$\int_0^1 \frac{dt}{X(x,t)} = \tau(x)$.

the spectrum of $T_a f$ is contained in $\{z : |z| < 1\}$; a is *repelling* if the spectrum of $T_a f$ is contained in $\{z : |z| > 1\}$.

The study of a fixed point is a *local problem*, i.e., one which can be studied in a small local chart. (By contrast, *global problems* involve the way in which the various local charts are attached to each other.) Let thus (U, φ, E) be a C^r chart of M around a with $r \geqslant 1$. We may assume that φa is the origin 0 of E. The map $\tilde{f} = \varphi^{-1} \circ f \circ \varphi$ is then defined "near 0," i.e., in a neighborhood of 0 in E (there is no need to say exactly which neighborhood, because our problem is local). It is equivalent to study f around a or $\tilde{f}$ around 0, and $D_0 \tilde{f}$ is linearly conjugate to $T_a f$.[14]

The hyperbolicity of the linear operator $A = D_0 \tilde{f}$ permits its decomposition into two parts: one with its spectrum inside the unit circle, the other with its spectrum outside the unit circle. More precisely, there exist closed subspaces E^- and E^+ of E with the following properties.

(a) E^- and E^+ are complementary, i.e., $E^- \cap E^+ = \{0\}$, $E^- + E^+ = E$.

(b) $AE^- \subset E^-$ and the spectrum of the restriction A_- of A to E^- is contained in $\{z : |z| < 1\}$.

(c) $AE^+ = E^+$ and the spectrum of the restriction A_+ of A to E^+ is contained in $\{z : |z| > 1\}$.

[(a), (b), (c) are easily proven in finite dimension. For the general case, see Appendix A.5.] The operators A_- and $(A_+)^{-1}$ have spectral radii less than 1. This implies (see Appendix A.5) that there is $\rho < 1$ such that

$$\lim_{n \to \infty} \|(A_-)^n\|^{1/n} < \rho, \qquad \lim_{n \to \infty} \|(A_+)^{-n}\|^{1/n} < \rho.$$

In view of this, there exist new norms $||| \cdot |||_{\pm}$ on $E^{\pm}$ (equivalent to the old norms) such that A_- and $(A_+)^{-1}$ are contractions. In fact, we obtain $|||A_-|||_- < \rho$, $|||(A_+)^{-1}|||_+ < \rho$ if we take

$$|||X|||_- = \sum_{n=0}^{\infty} \rho^{-n} \|(A_-)^n X\|,$$

$$|||Y|||_+ = \sum_{n=0}^{\infty} \rho^{-n} \|(A_+)^{-n} Y\|.$$

[14]This means that $D_0 \tilde{f} = L^{-1} T_a f L$ for some continuous linear map L with continuous inverse (see Equation (4.1)). In particular, $D_0 \tilde{f}$ and $T_a f$ have the same spectrum.

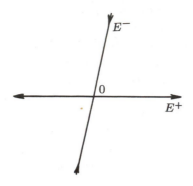

F<small>IG</small>. 8.   Contracting and expanding subspaces conventionally marked by arrows.

Finally, we may define $||| \cdot |||$ on E by $|||X + Y||| = \max\{|||X|||_-,$ $|||Y|||_+\}$ if $X \in E^-$, $Y \in E^+$.

Up to a permissible change of norm on $T_a M$, we have thus decomposed $A = D_0 \tilde{f}$ into a linear contraction A_- on E^-, and an invertible operator A_+ on E^+ that has a contracting inverse. The spaces E^- and E^+ are uniquely determined by

(5.1) $E^- = \{X : A^n X \to 0\},$

(5.2)

$$E^+ = \{Y_0 : \exists (Y_n)_{n \geqslant 0} \quad \text{with} \quad AY_{n+1} = Y_n \quad \text{and} \quad Y_n \to 0.$$

The subspaces E^-, E^+ of E are the *contracting* and *expanding* subspaces, respectively, for the hyperbolic operator A (see Fig. 8). One also says that $V_a^- = (T\varphi^{-1})E^-$, $V_a^+ = (T\varphi^{-1})E^+$ are the contracting and expanding subspaces (of $T_a M$) for f at a. If f is a diffeomorphism, replacing f by f^{-1} interchanges contracting and expanding subspaces.

If the fixed point a is attracting, then $E^+ = \{0\}$. Since f is smooth, the function $x \mapsto D_x \tilde{f}$ is continuous and we may assume that $|||D_x \tilde{f}||| < \beta < 1$ for $|||x||| < \varepsilon$, with $\varepsilon > 0$. If x is in the ball $E_0(\varepsilon)$ of radius ε centered at 0, we have $|||\tilde{f}x||| \leqslant \beta|||x|||$. Hence, $\tilde{f}^n x \to 0$ when $n \to \infty$. Therefore, $f^n x \to a$ for x in the neighborhood $\varphi^{-1}E_0(\varepsilon)$ of a. This justifies calling a an attracting fixed point. In fact, we see that the attraction is *exponentially fast*.

If a is a repelling fixed point, then $E^- = \{0\}$, and f is locally invertible near a by the inverse function theorem B.3.1 (f is a *local diffeomorphism*). The fixed point a is attracting for the map f^{-1} (locally defined), and this justifies calling a a repelling fixed point for f.

5.2. Theorem (Grobman–Hartman for maps). *Let E be a Banach space, $\tilde{U}$ a neighborhood of 0 in E, and $\tilde{f} : \tilde{U} \mapsto E$ a C^1 map. If 0 is a hyperbolic fixed point, and if $D_0\tilde{f}$ is invertible, then $\tilde{f}$ is topologically conjugate to $D_0\tilde{f}$ near 0. This means that there is an open set $V \ni 0$ and a homeomorphism h of V to a neighborhood of 0 in $\tilde{U}$ such that $h0 = 0$, and*

$$h(D_0\tilde{f}) = \tilde{f}h.$$

Notice that the assumed invertibility of $D_0\tilde{f}$ and the inverse function theorem imply that $\tilde{f}$ is a local diffeomorphism. Note also that the map h is in general not smooth (even if $\tilde{f}$ is C^∞ or analytic). Differentiability can be obtained at the expense of making further assumptions (see Section 5.8).

The proof of the Grobman–Hartman theorem uses the fact that $\tilde{f}$ is C^1 close to $D_0\tilde{f}$ near 0 (i.e., $\tilde{f} - D_0\tilde{f}$ and its first derivative have arbitrarily small norm in $E_0(\varepsilon)$ if ε is small enough). It is then possible to obtain g by a contraction mapping argument.[15]

The Grobman–Hartman theorem is useful for the intuitive picture it gives of a diffeomorphism near a fixed point (a local diffeomorphism f near a hyperbolic fixed point a is topologically conjugate to the derivative D_af). We shall, however, find that the invariant manifold theorems discussed in later sections are perhaps more valuable tools in describing dynamics.

5.3. Fixed Points of a Semiflow

Let a be a fixed point for the semiflow (f^t). We assume here that the maps f^t are smooth, that $(x, t) \mapsto f^t x$ is continuous on $M \times \{t : t \geqslant 0\}$, and that $t \mapsto T_a f^t$ is continuous on $\{t : t > 0\}$. We say that a is *hyperbolic* (resp. *attracting, repelling*) as a fixed point of (f^t) if it is hyperbolic (resp. attracting, repelling) as a fixed point of the time one map f^1.

Since f^1 is hyperbolic, we may apply the notation and results of Section 5.1 to $f = f^1$. We also write $\tilde{f}^t = \varphi^{-1} \circ f^t \circ \varphi$ and $A^t = D_0\tilde{f}^t$; (A^t) is then a semigroup of linear operators, and $t \mapsto A^t$ is continuous for $t > 0$. In view of (5.1) and (5.2), $A^t E^- \subset E^-$ and $A^t E^+ = E^+$ for all $t > 0$. The restrictions $A^t_\pm$ of A^t to $E^\pm$ define semigroups (A^t_-),

---

[15] The origins of Theorem 5.2 are in Grobman [1] and Hartman [1]. For a proof using the contraction mapping theorem, see Palis [1] or Pugh [2]. For a geometric proof using the inclination lemma, see Problem 6.

(A_+^t) such that $t \mapsto A_-^t$, A_+^t are continuous for $t > 0$. In particular, $|||A_-^\tau||| < \beta$, $|||(A_+^\tau)^{-1}||| < \beta$ for τ close to 1, with $\beta < 1$. For all sufficiently large t we may write $t = n\tau$ with $\tau \geqslant 1$ so that $|||A_-^t||| < \beta^t$, $|||(A_+^t)^{-1}||| < \beta^t$. Therefore, for all $t > 0$, the spectral radii of A_-^t, $(A_+^t)^{-1}$ are

$$\lim_{n \to \infty} |||(A_-^t)^n|||^{1/n} < \beta^t, \qquad \lim_{n \to \infty} |||(A_+^t)^{-n}||| < \beta^t.$$

In particular, A^t is hyperbolic with the same subspaces E^-, E^+ as A^1. Conversely, the hyperbolicity of A^t for any $t > 0$ implies the hyperbolicity of A^1 (use the semigroup (B^τ) with $B^\tau = A^{\tau t}$).

In conclusion, the fixed point a is hyperbolic (and, in particular, attracting or repelling) for the semiflow (f^t) if and only if it has this property for one, and therefore all, f^t, $t > 0$. The same contracting and expanding subspaces $V_a^- = (T\varphi^{-1})E^-$, $V_a^+ = (T\varphi^{-1})E^+$ are obtained for all $t > 0$, and are called contracting and expanding subspaces for (f^t). One sees easily that attracting and repelling fixed points deserve their names as in the case of maps.

If (f^t) is the flow obtained by integration of a C^1 vector field X, the fixed points of (f^t) are the points where X vanishes (*critical points* of the vector field). At a fixed point a, we have

$$D_a f_\mu^t = \exp t D_a X_\mu.$$

Therefore, a is hyperbolic (resp. attracting, repelling) if and only if the spectrum of $D_a X$ is disjoint from the imaginary axis (resp. to its left, to its right).

5.4. Theorem (Grobman–Hartman for flows).[16] *Let E be a Banach space and X a C^1 vector field defined in a neighborhood of the origin 0 of E. Suppose that 0 is a hyperbolic fixed point for the local flow (f^t) defined by X. There is then a homeomorphism h from a neighborhood of 0 to a neighborhood of 0 in E such that $h0 = 0$ and*

$$h \circ e^{t(D_0 X)} = f^t \circ h$$

(when both sides are defined).

The local flow defined by X is thus topologically conjugate to the flow defined by $D_0 X$ near 0. As for maps, differentiability can be obtained at the expense of making further assumptions (see Section 5.8).

---

[16] See Palis [1] or Pugh [2] for a proof.

5.5. *Periodic Orbits for a Map*

Let a be a periodic point of period n for the smooth map f. We say that the periodic orbit $\{f^k a\}$ is *hyperbolic* (resp. *attracting*) if a is hyperbolic (resp. attracting) as a fixed point of f^n. We say that the periodic orbit $\{f^k a\}$ is *repelling* if a is repelling as a fixed point of f^n, and if f is a local diffeomorphism near each $f^k a$, $k = 0, \dots, n-1$.

For $0 < k < n$, let $A = T_a f^k$, $B = T_{f^k a} f^{n-k}$. If $T_a f^n = BA$ is hyperbolic (or attracting), $T_{f^k a} f^n = AB$ is also hyperbolic (or attracting) because, if $z \neq 0$, the existence of $(1 - z^{-1}BA)^{-1} = C$ implies the existence of $(1 - z^{-1}AB)^{-1} = 1 + z^{-1}ACB$. The contracting and expanding subspaces V_a^-, V_a^+ for f^n at a are also called contracting and expanding subspaces for f at the periodic point a. We have $(T_a f)V_a^- \subset V_{fa}^-$ and $(T_a f)V_a^+ = V_{fa}^+$ (use (5.1), (5.2) of Section 5.1). The periodic orbit is attracting (resp. repelling) if $V_{f^k a}^+ = \{0\}$ (resp. $V_{f^k a}^- = \{0\}$) for $k = 0, \dots, k-1$. The orbit is attracting (repelling) if and only if each $f^k a$ is attracting (repelling) for f^n. As for fixed points, the names attracting and repelling are deserved.

5.6. *Periodic Orbits for a Semiflow*

Let a be a periodic point of period $T(a) > 0$ for the semiflow (f^t), and call Π the periodic orbit through a. We assume that $(x, t) \mapsto f^t x$ is continuous on $M \times \{t : t \geqslant 0\}$ and C^r in (neighborhood of Π) $\times \{t : t > 0\}$ with $r \geqslant 1$. We say that the periodic orbit Π is *hyperbolic* (resp. *attracting*) if a is hyperbolic (resp. attracting) as a fixed point of the Poincaré map P corresponding to a local section Σ (see Proposition 4.1). We say that the periodic orbit Π is *repelling* if a is a repelling fixed point for P, and the maps f^t are local diffeomorphisms near a, for $t > 0$. Notice that a is *not* a hyperbolic fixed point for $f^{T(a)}$!

We now compare P to the Poincaré map Q corresponding to a cross-section Σ' through a point b of the periodic orbit. We may take $b \neq a$ (if not, compare P, Q with the Poincaré map for a cross-section through $c \neq a$). There are maps $g : \Sigma \mapsto \Sigma'$, $g' : \Sigma' \mapsto \Sigma$ (see Fig. 9) such that $g' \circ g$ is P and $g \circ g'$ is Q, if one is close enough to a in Σ or b in Σ'. The existence of these maps and the fact that they are C^r are *obvious* in the sense discussed just before the statement of Proposition 4.1. Let $A = T_a g$, $B = T_b g'$. The spectrum of AB is the same as that of BA up to the possible exception of 0 (see Section 5.5) so that Q is hyperbolic (or attracting) at b if and only if P is hyperbolic (or attracting) at a.

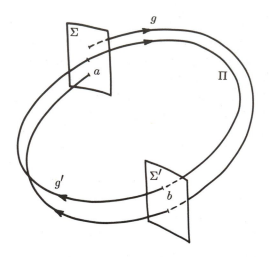

FIG. 9.   Comparing two Poincaré maps.

Let $V_{\Pi a}^-$ be the subspace of $T_a M$ spanned by the contracting subspace for P together with the vector X giving the direction of the semiflow (Section 4). Similarly, let $V_{\Pi a}^+$ be the subspace spanned by the expanding subspace for P and by X. Every vector $Y \in T_a M$ has a unique decomposition into a vector in $T_a \Sigma$ and a multiple of X; this gives the characterizations

$$V_{\Pi a}^- = \{Y \in T_a M : \|(T_a f^t)Y\| \text{ is bounded for } t \geqslant 1\},$$
$$V_{\Pi a}^+ = \{Y_0 \in T_a M : \exists (Y_s)_{s \in \mathbf{R}} \text{ with } (T_a f^t)Y_s = Y_{t+s},$$
$$\text{and } \|Y_s\| \text{ is bounded for } s \leqslant -1\},$$

and also

$$V_{\Pi a}^- \cap V_{\Pi a}^+ = \mathbf{R} X.$$

The space $V_{\Pi a}^-$ is called contracting space, and the space $V_{\Pi a}^+$ is called expanding space for the semiflow (f^t) at the point a of the hyperbolic periodic orbit $\{f^t a\}$.

5.7. Sinks, Sources, and Saddles

A hyperbolic fixed point or periodic orbit is also called a *sink* if it is attracting, a *source* if it is repelling; otherwise it is a *saddle* (or of saddle type).

5.8. Smooth Linearization

The Grobman–Hartman theorem (for maps of flows) is satisfactory in providing a local *topological* model for dynamics. It is often relevant to ask for *differentiable* models, requiring the conjugacy h as above to be differentiable. The following refinements of Theorems 5.2 and 5.4 assume that the eigenvalues of $D_0\tilde{f}$ (maps) or D_0X (flows) satisfy certain *nonresonance* conditions besides hyperbolicity.

(Maps, h of class C^∞). *If E is finite dimensional and $\tilde{f}$ of class C^∞, one can take for h a C^∞ diffeomorphism, provided the eigenvalues $\lambda_1, \ldots, \lambda_m$ of $D_0\tilde{f}$ satisfy the conditions $\lambda_i \neq \lambda_1^{k_1} \cdots \lambda_m^{k_m}$ for all i and $k_1, \ldots, k_m \geq 0$ such that $k_1 + \cdots + k_m \geq 2$.*

(Flows, h of class C^∞). *If E is finite dimensional and X of class C^∞, one can take for h a C^∞ diffeomorphism, provided the eigenvalues $\chi_1, \ldots, \chi_m$ of D_0X satisfy the conditions $\chi_i \neq k_1\chi_1 + \cdots + k_m\chi_m$ for all i and $k_1, \ldots, k_m \geq 0$ such that $k_1 + \cdots + k_m \geq 2$.*

The nonresonance conditions in these statements ensure that there is a formal power series conjugating $\tilde{f}$ to $D_0\tilde{f}$ near 0 for maps, and similarly for flows.

(Maps, h of class C^1). *If E is finite dimensional and $\tilde{f}$ of class $C^{(1,1)}$, one can take for h a C^1 diffeomorphism, provided the eigenvalues of $D_0\tilde{f}$ satisfy $|\lambda_i| \neq |\lambda_j\lambda_k|$ when $|\lambda_j| < 1 < |\lambda_k|$.*

(Flows, h of class C^1). *If E is finite dimensional and X of class $C^{(1,1)}$, one can take for h a C^1 diffeomorphism provided the eigenvalues of D_0X satisfy $\operatorname{Re} X_i \neq \operatorname{Re}\chi_j + \operatorname{Re}\chi_k$ when $\operatorname{Re}\chi_j < 0 < \operatorname{Re}\chi_k$.*

There are more complicated criteria for the existence of C^r conjugacies for other values of r. Extensions to nonhyperbolic cases are also possible.[17]

---

[17] The above C^∞ statements are due to Sternberg [1], [2], following the treatment of C^ω flows by Siegel [1]. For C^r statements with r finite, see Sternberg [2], Nelson [1], Takens [1], [2] (who considers nonhyperbolic situations), and Belitskii [1]. The above C^1 statement for maps is in Belitskii [1]. The flow case is obtained by a standard argument (Lemma 4 of Sternberg [1]).

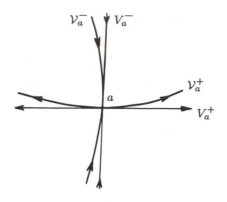

FIG. 10.   Stable and unstable manifolds of a hyperbolic fixed point a.

6. Stable and Unstable Manifolds

Consider again a hyperbolic fixed point a for a local smooth map f. In the previous section we introduced a contracting linear space V_a^- and an expanding linear space V_a^+. There are also nonlinear contracting and expanding spaces called the (local) stable manifold $\mathcal{V}_a^-$ and the (local) unstable manifold $\mathcal{V}_a^+$. That these are local concepts means that they are defined in a suitably small neighborhood U of a in the *ambient* manifold M. The *stable manifold* $\mathcal{V}_a^-$ consists of those x such that $f^n x \to a$ when $n \to +\infty$ while $f^n x$ remains in U for all $n \geq 0$. (For small U, it suffices in fact to require that the forward orbit $\{f^n x : n \geq 0\}$ stays in U.) The *unstable manifold* $\mathcal{V}_a^+$ consists of those x_0 for which there is a sequence (x_n) in U with $f x_{n+1} = x_n$ and $x_n \to a$ (in fact, $x_n \to a$ again comes for free if U is small enough).

If f is a local diffeomorphism, the replacement of f by f^{-1} interchanges $\mathcal{V}_a^-$ and $\mathcal{V}_a^+$. (Note also that in that case, $\mathcal{V}_a^\pm$ correspond to $V_a^\pm$ by the homeomorphism introduced in the Grobman–Hartman Theorem 5.2.)

The next theorem will show that $\mathcal{V}_a^-$ and $\mathcal{V}_a^+$ are smooth manifolds tangent to V_a^- and V_a^+, respectively, at a (see Fig. 10). To formulate this result, we relax the condition of hyperbolicity and make the somewhat weaker assumption that a is a ρ-*pseudohyperbolic* fixed point of f. This means that the spectrum of $T_a f$ is disjoint from the circle $\{z \in \mathbf{C} : |z| = \rho\}$. According to Appendix A.5, we may associate closed subspaces V_a' and V_a'' of $T_a M$ to the parts of the spectrum of $T_a f$ in $\{z : |z| < a\}$ and $\{z : |z| > a\}$, respectively. For $\rho \leq 1$, V_a' is a contracting space, and for $\rho \geqslant 1$, V_A'' is an expanding space. More or less contracting or expanding spaces may be obtained for different ρ's. Of course, for $\rho = 1$, $V_a' = V_a^-$ and $V_a'' = V_a^+$. If $\rho \leqslant 1$, a local manifold V_a' tangent to V_a' will be defined; if $\rho \geqslant 1$, a local manifold V_a'' tangent to V_a'' will be defined. We call V_a' a *stable* manifold and V_a'' an *unstable* manifold. For $\rho = 1$ we recover the stable manifold $V_a^- = V_a'$ and the unstable manifold $V_a^+ = V_a''$ of the hyperbolic fixed point a.

As in Sections 5.1 and 5.2, we use a chart (U, φ, E) such that $\varphi a = 0$ and write $\tilde{f} = \varphi^{-1} \circ f \circ \varphi$.

6.1. Theorem (Stable and unstable manifold theorem). *Let E be a Banach space, $\tilde{U}$ a neighborhood of 0 in E, and $\tilde{f} : \tilde{U} \mapsto E$ a C^r map, $r \geqslant 1$. We assume that $\tilde{f}0 = 0$ and that the spectrum of $D_0\tilde{f}$ is disjoint from the circle $|z| = \rho$ (ρ-pseudohyperbolicity). We call E', E'' the subspaces of E associated with the parts of the spectrum of $D_0\tilde{f}$ inside and outside, respectively, of the circle $|z| = \rho$. Also, let A', A'' be the operators on E', E'' induced by $D_0\tilde{f}$. Going to equivalent norms $||| \cdot |||$, we may (and do) assume that $|||A'||| < \rho$, $|||A''^{-1}||| < \rho^{-1}$, and*

$$E = E'' \oplus E'.$$

(Stable manifold). *If $\rho \leqslant 1$, then, for sufficiently small $R > 0$, the set*

$$(6.1) \qquad V' = \bigcap_{n \geqslant 0} \tilde{f}^{-n} E_0(\rho^n R)$$

is the graph of a C^r map $\varphi' : E_0'(R) \mapsto E_0''(R)$ with $\varphi'(0) = 0$, $D_0\varphi' = 0$.

(Unstable manifold). *If $\rho \geqslant 1$, then for sufficiently small $R > 0$, the set*

$$(6.2) \qquad V'' = \bigcap_{n \geqslant 0} \tilde{f}^n \bigcap_{m=0}^{n} \tilde{f}^{-m} E_0(\rho^{m-n} R),$$

which coincides with

$$\{x_0 : \exists (x_k)_{k \leqslant 0} \text{ with } \tilde{f}x_{k-1} = x_k \text{ and } x_k \in E_0(\rho^k R)\},$$

is the graph of a C^r map $\varphi'' : E_0''(R) \to E_0'(R)$ with $\varphi''(0) = 0$, $D_0\varphi'' = 0$.

The manifolds V', V'' depend continuously on $\tilde{f}$ for the C^r topologies.[18]

Notice that the value of ρ in the theorem may be varied over an interval without changing E', E''. If R is small enough, V', V'' are also not changed. [Let $\tilde{\rho} < \rho \leqslant 1$, then $\tilde{V}' \subset V'$, but $\tilde{V}'$ and V' are both graphs of functions $E_0'(R) \mapsto E_0''(R)$, and therefore $\tilde{V}' = V'$; similarly for V'']. The manifolds V', V'' are asdifferentiable as the map f. They are tangent at 0to E' and E'', respectively, and satisfy

$$(f^{-1}V') \cap E_0(R) = V', \qquad (fV'') \cap E_0(R) = V''.$$

Conversely, if for small R, a C^1 manifold $V \subset E_0(R)$ is tangent to E' (resp. E'') at 0 and satisfiesthe *local invariance*

$$(fV) \cap E_0(R) \subset V,$$

then $V = V'$ (resp. $V = V''$). [For small R, $V \subset V'$ bydefinition of V', and since V, V' are both graphs of functions $E_0'(R) \mapsto E_0''(R)$,we have $V = V'$; similarly for V''.]

6.2. On the Proof of the Stable and Unstable Manifold Theorems

Theorem 6.1 has a number of variants and generalizations[19] and there are several approaches to the proof. Whereas the details are long and will not be reproduced here, the ideas are simple and well worth understanding.

We first describe the *graph transform method*, originating with Hadamard and later developed extensively.[20] For simplicity, we assume that $D_0\tilde{f}$ is invertible and that $\rho = 1$. Using the notation of Appendix

---

[18]The C^r topology is defined for $r < \infty$ by a norm $\|\cdot\|_r$, and for $r = \infty$ by a sequence $\|\cdot\|_r$, $r = 0, 1, \ldots$ (see Appendix B.1 where the real analytic case is also handled).

[19]The hyperbolicity assumption may be dropped to obtain the center stable and center unstable manifold theorem 7.1. Another possibility is to replace the fixed point 0 by a compact manifold and to assume normal hyperbolicity (Theorem 14.2). One may also generalize the notion of a hyperbolic periodic orbit and obtain stable and unstable manifolds for points of a more general hyperbolic set (Section 15.2), or make no hyperbolicity assumption and use a measure-theoretic approach ("Pesin theory," not discussed in this volume, see Pesin [2], [3], Ruelle [3], [5], Mañé [1]).

[20]The monograph by Hirsch, Pugh, and Shub [1] is a standard reference and contains many general results, but is tough reading. A more accessible version of the stable manifold theorem is available in Shub [2] (Chapter 5).

B.1, we define[21]

$$\mathcal{L}_1 = \{\psi \in C^{(0,1)}(E_0''(R), E_0'(R)) : \psi(0) = 0 \quad \text{and} \quad \|\psi\|_{(0,1)} \leqslant 1\}.$$

If $\psi \in \mathcal{L}_1$, the graph $\Gamma(\psi)$ is a subset of $E_0''(R) \times E_0'(R) = E_0(R)$.[22] One can show that the restriction to $E_0(R)$ of $\tilde{f}\Gamma(\psi)$ is again the graph of a function in $\mathcal{L}_1$ (see Fig. 11). We denote this new function by $\tilde{f}_\# \psi$, so that $\tilde{f}\Gamma(\psi) \cap E_0(R) = \Gamma(\tilde{f}_\# \psi)$. The functional map $\tilde{f}_\#$ is the *graph transformation* corresponding to $\tilde{f}$. Since $\tilde{f}$ is a small perturbation of $D_0\tilde{f}$, the map $\tilde{f}_\#$ stretches the graph of a function in the unstable direction E'' and compresses it in the stable direction E'. The geometric properties of $\tilde{f}$ are thus reflected in analytic properties of $\tilde{f}_\#$, and the graph transformation turns out to be an excellent tool for the study of the unstable manifold. In fact, $\tilde{f}_\#$ is a contraction on $\mathcal{L}_1$ equipped with the *uniform* norm: $\|\psi\|_0 = \sup_x |\psi(x)|$. The fixed point of $\tilde{f}_\#$ will be the function φ'' defining the unstable manifold. The fact that $\tilde{f}_\#$ stretches in the unstable direction and compresses in the stable direction tends to decrease large derivatives. This explains the smoothness of the fixed point φ'' and why $D_0\varphi'' = 0$.[23] If $(x'', x') \in E_0''(R) \times E_0'(R)$, R small, the distance of (x'', x') to $(x'', \varphi''(x''))$ is multiplied by a factor < 1 when we apply $\tilde{f}$. We may thus estimate the distance from $\tilde{f}^k(x'', x')$ to the graph of φ'' when $k = 1, 2, \ldots$. This permits the identification $\mathcal{V}'' = \bigcap_{n \geqslant 0} \tilde{f}^n E_0(R)$. To fill in the details of the proof of Theorem 6.1 requires some work.[24]

While Hadamard's method applies most naturally to the unstable manifold, Perron proved the smoothness of the stable manifold by starting from the definition $\{x : \tilde{f}^n x \to 0\}$. The stable and unstable manifold theorem can also be proven by a clever application of the implicit function theorem[25] or by other techniques. In the analytic finite dimensional case, 1-dimensional unstable manifolds can be obtained in

---

[21] $\mathcal{L}_1$ is the set of functions $E_0''(R) \longmapsto E_0'(R)$ vanishing at 0 and with Lipschitz constant $\leqslant 1$, which means that $\|\psi(x) - \psi(y)\| \leqslant \|x - y\|$ for all x, y.

[22] Here we identify $E'' \times E'$ with $E'' \oplus E'$. See Appendix A.4.

[23] Technically, one uses the *invariant section theorem* of Hirsch, Pugh, and Shub [1]§3 (see also Shub [2], théorème 5.18).

[24] Theorem 6.1 is proven in Hirsch, Pugh, and Shub [1] (Theorems 5.1, 5.4) when r is an integer $\geqslant 1$, or $r = \infty$. The case $r = (r, \alpha)$ is an easy variation. To handle the C^ω case, it suffices to use a holomorphic extension of $\tilde{f}$ to a neighborhood of 0 in $E_{\mathbf{C}}$, apply the C^1 version of the theorem to this extension, and then restrict to E.

[25] This is Irwin's proof. See Irwin [1], and also Shub [2] Appendix 5.2.

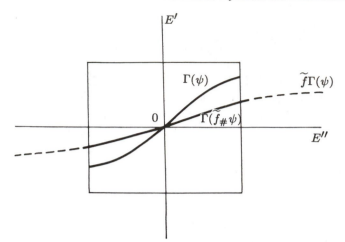

FIG. 11.   The graph transform $\tilde{f}_{\#}\psi$ of a function $\psi : E_0''(R) \mapsto E_0'(R)$.

the form of power series with the coefficients determined by substitution. This was studied by Poincaré[26] and has often been rediscovered since.

6.3. Stable and Unstable Manifolds for Fixed Points of a Semiflow and for Periodic Orbits

For simplicity, we discuss only the hyperbolic situation. The ρ-pseudohyperbolic case may be handled similarly.

Let a be a hyperbolic fixed point for the semiflow (f^t). According to Section 5.3, a is nothing other than a hyperbolic fixed point for the time one map f^1. If V_a^- is a (local) stable manifold with respect to f^1 and $x \in V_a^-$, then $f^t x \to a$ when $t \to +\infty$ and $f^t x$ remains in a small neighborhood U of a. It is thus natural to say that V_a^- is a (local) stable manifold with respect to (f^t). Similarly, we take as a (local) unstable manifold V_a^+ with respect to (f^t), a (local) unstable manifold V_a^+ with respect to f^1. V_a^+ consists of points x_0 for which there is a family $(x_s)_{s \geqslant 0}$ with $f^t x_s = x_{s-t}$ if $t \leqslant s$,

---

[26] See Poincaré [1], [2] (these papers are reproduced in Poincaré [5]). Poincaré shows that if λ is a simple eigenvalue of $D_0 \tilde{f}$ such that $|\lambda| > 1$ and no power λ^n (n positive integer) is again an eigenvalue, then there is an analytic invariant manifold tangent to the eigenvector of $D_0 \tilde{f}$ corresponding to λ. If $\tilde{f}$ is a polynomial, this unstable manifold is defined parametrically in terms of entire functions.

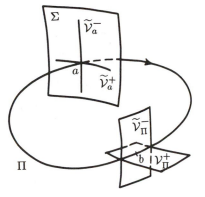

FIG. 12.   Stable and unstable manifolds of a periodic orbit Π for a semiflow.

$x_t \to a$ when $t \to \infty$, and x_s remains in a small neighborhood U of a.

One could also define $V_a^- = \{x : f^t x \in U \text{ for all } t \geqslant 0\}$ and similarly for V_a^+. This new definition of local stable and unstable manifolds agrees with the previous one locally, namely, sufficiently near a, i.e., if one intersects with a sufficiently small neighborhood $\widetilde{U}$ of a.

Let $\{f^k a\}$ be a hyperbolic periodic orbit for a map f with period n. We define the stable and unstable manifolds for f^n as the stable and unstable manifolds V_a^- and V_a^+ through the periodic point a. One also calls $\bigcup_{k=0}^{n-1} V_{f^k a}^{\pm}$ a (un)stable manifold for the periodic orbit. The properties of these stable and unstable manifolds can be read off from Theorem 6.1.

Let $\Pi = \{f^t a\}$ be a hyperbolic periodic orbit of period $T(a) > 0$ for the semiflow (f^t), and U a small neighborhood of Π. We define a stable manifold of Π by

$$V_{\Pi}^- = \{x : f^t x \in U \text{ for all } t \geqslant 0\},$$

and an unstable manifold by

$$V_{\Pi}^+ = \{x : \exists\, (x_s)_{s \geqslant 0} \text{ with } f^t x_s = x_{s-t} \text{ for } t \leqslant s$$
$$\text{and } x_s \in U \text{ for all } s \geqslant 0\}.$$

We expect that V_{Π}^- and V_{Π}^+ are C^r manifolds (near Π) tangent to $V_{\Pi b}^-$ and $V_{\Pi b}^+$, respectively, at $b \in \Pi$ (see Section 5.6). To see that this is indeed the case, we shall do a little exercise using the Poincaré map and different forms of the implicit function theorem. Let $\widetilde{V}_a^{\pm}$, $\widetilde{\mathcal{V}}_a^{\pm}$ be the expanding and contracting spaces and the stable and unstable manifolds for the Poincaré map P corresponding to a local section Σ through a (see Fig. 12). A point x close enough to Π is in $\mathcal{V}_\Pi^-$ if and only if the first intersection y of the orbit through x with Σ is in $\widetilde{\mathcal{V}}_a^-$. The map $x \mapsto y$ is transversal to $\widetilde{\mathcal{V}}_a^-$ at b because $\widetilde{V}_a^+$ is in the image of the tangent map. Therefore, $\mathcal{V}_\Pi^-$ is $C^{\mathbf{r}}$ near b and tangent to $V_{\Pi b}^-$ (see Appendix B.3.5). A point x close enough to Π is in $\mathcal{V}_\Pi^+$ if and only if it is of the form $f^t y$ with $y \in \widetilde{\mathcal{V}}_a^+$. If $b = f^\tau a$, the map $(y, t) \mapsto f^t y$ is an immersion of $\widetilde{\mathcal{V}}_a^+ \times \mathbf{R}$ at (a, τ), and $\mathcal{V}_\Pi^+$ is therefore $C^{\mathbf{r}}$ near b and tangent to $V_{\Pi b}^+$ (see Appendix B.3.4).

In Section 14 below, we shall define (un)stable manifolds $\mathcal{V}_x^{\pm}$ of points $x \in \Pi$ (*strong* (un)stable manifolds), and we shall have $\mathcal{V}_\Pi^{\pm} = \bigcup_{x \in \Pi} \mathcal{V}_x^{\pm}$.

7. Center Manifolds

The study of fixed points and periodic orbits becomes more delicate when they are not hyperbolic. One is, however, forced to consider non-hyperbolic behavior in the theory of bifurcations (see Part 2). The following analog of Theorem 6.1 applies to nonhyperbolic fixed points. Note the assumption $\mathbf{r} < \infty$, and the condition that E' or E'' has the $C^{\mathbf{r}}$ extension property (see Appendix B.1; the $C^{\mathbf{r}}$ extension property always holds for finite-dimensional spaces).

7.1. Theorem (Center stable and center unstable manifold theorem). *Let E be a Banach space, $\widetilde{U}$ a neighborhood of 0 in E, and $\tilde{f} : \widetilde{U} \mapsto E$ a $C^{\mathbf{r}}$ map, $1 \leqslant \mathbf{r} < \infty$. We assume that $\tilde{f}0 = 0$ and that the spectrum of $D_0\tilde{f}$ is disjoint from the circle $|z| = \rho$. We call E', E'' the subspaces of E associated with the parts of the spectrum of $D_0\tilde{f}$ inside and outside, respectively, of the circle $|z| = \rho$, and let A', A'' be the operators on E', E'' induced by $D_0\tilde{f}$. Going to equivalent norms $||| \cdot |||$, we may assume that $|||A'||| < \rho$, $|||A''^{-1}||| < \rho^{-1}$, and*

$$E = E'' \oplus E'.$$

We write $|\mathbf{r}| = r$ if $\mathbf{r}$ is the integer r, and $|\mathbf{r}| = r + \alpha$ if $\mathbf{r} = (r, \alpha)$.
(Center stable manifold). *Suppose that $\rho \geqslant 1$, $|||A''^{-1}||| < |||A'|||^{-|\mathbf{r}|}$, and let E' have the $C^{\mathbf{r}}$ extension property. Then, for sufficiently small $R > 0$, there is a $C^{\mathbf{r}}$ map $\varphi' : E_0'(R) \mapsto E_0''(R)$ with $\varphi'(0) = 0$,*

$D_0\varphi' = 0$, *such that the graph* $\mathcal{V}'$ *of* φ' *satisfies the local invariance properties*

$$(f\mathcal{V}') \cap E_0(R) \subset \mathcal{V}', \qquad (f^{-1}\mathcal{V}') \cap E_0(R) \subset \mathcal{V}'.$$

Furthermore, if $x_n \in \bigcap_{k=0}^{n} f^{-k} E_0(R)$, *the distance of* x_n *to* $\mathcal{V}'$ *tends to* 0 *when* $n \to \infty$ ($\mathcal{V}'$ *is locally attracting for* f^{-1}).

(Center unstable manifold). *Suppose that* $\rho \leqslant 1$, $|||A'||| < |||A''^{-1}|||^{-|\mathbf{r}|}$, *and let* E'' *have the* $C^{\mathbf{r}}$ *extension property. Then, for sufficiently small* $R > 0$, *there is a* $C^{\mathbf{r}}$ *map* $\varphi'' : E''(R) \mapsto E'(R)$ *with* $\varphi''(0) = 0$, $D_0\varphi'' = 0$ *such that the graph* $\mathcal{V}''$ *of* φ'' *is locally invariant:*

$$(f\mathcal{V}'') \cap E_0(R) \subset \mathcal{V}''.$$

Furthermore, if $x_n \in \bigcap_{k=0}^{n} f^{-k} E_0(R)$, *the distance of* $f^n x_n$ *to* $\mathcal{V}''$ *tends to* 0 *when* $n \to \infty$ ($\mathcal{V}''$ *is locally attracting for* f).

The manifolds $\mathcal{V}'$, $\mathcal{V}''$ *are not claimed to be unique, but may be chosen to depend continuously on* $\tilde{f}$ *for the* $C^{\mathbf{r}}$ *topologies.*

If the spectrum of $D_a f$ is the union of one part in $\{z : |z| \leqslant 1\}$ and one part in $\{z : |z| > \rho\}$, $\rho > 1$, the condition $|||A''^{-1}||| < |||A'|||^{-|\mathbf{r}|}$ can always be satisfied by a suitable choice of $||| \cdot |||$ if $\mathbf{r} < \infty$. If E' has the $C^{\mathbf{r}}$ extension property, a manifold $\mathcal{V}'_a$ is obtained, and is called a *center stable manifold*. We denote this manifold by $\mathcal{V}_a^{0-}$ and denote the corresponding tangent space at a by V_a^{0-}. Note that in the present situation, an unstable manifold $\mathcal{V}_a^+$ is also defined by Theorem 6.1.

If the spectrum of $D_a f$ is the union of one part in $\{z : |z| < \rho\}$, $\rho < 1$, and one part in $\{z : |z| \geqslant 1\}$, the condition $|||A'||| < |||A''^{-1}|||^{-|\mathbf{r}|}$ can always be satisfied if $\mathbf{r} < \infty$. If E'' has the $C^{\mathbf{r}}$ extension property, a manifold $\mathcal{V}''_a$ is obtained, called a *center unstable manifold* of a. We denote this manifold by $\mathcal{V}_a^{0+}$ and denote the corresponding tangent space at a by V_a^{0+}. Note that in the present situation, a stable manifold $\mathcal{V}_a^-$ is also defined by Theorem 6.1.

In the more general situation of the theorem, one may still call $\mathcal{V}'_a$, $\mathcal{V}''_a$ center stable and center unstable manifolds in an extended sense.

Suppose that the conditions for the existence of $\mathcal{V}_a^{0-}$ and $\mathcal{V}_a^{0+}$ are simultaneously satisfied. The intersection $\mathcal{V}_a^0 = \mathcal{V}_a^{0-} \cap \mathcal{V}_a^{0+}$ is then called a *center manifold* of a. It is locally invariant, i.e., $(f\mathcal{V}_a^0) \cap E_0(R) \subset \mathcal{V}_a^0$ and tangent at a to the space V_a^0 associated with the part of the spectrum of $D_a f$, which is on the unit circle $\{z : |z| = 1\}$.

If V_a^- and V_a^{0-} are both defined, the stable manifold is contained in the center stable manifold. [In V_a^{0-} there is, by Theorem 6.1, a locally invariant manifold tangent to V_a^-. This manifold is V_a^- by uniqueness of the stable manifold.] Similarly, the unstable manifold is contained in the center unstable manifold.

7.2. Summary of (Locally) Invariant Manifolds of a Fixed Point

In the notation adopted, $-$ corresponds to the inside of the unit circle, and $+$ to the outside. The opposite notation is sometimes used, and it may be convenient to use instead of $-$, $+$, 0, the letters s (for stable, or contracting), u (for unstable, or expanding), and c (for center). The following table gives the relation between the various locally invariant manifolds, and parts of the spectrum of $D_a f$.

Manifold		Spectrum
Stable	$V^- = V^s$	$|z| < 1$
$\cap$		
Center stable	$V^{0-} = V^{cs}$	$|z| \leqslant 1$
$\cup$		
Center	$V^0 = V^c$	$|z| = 1$
$\cap$		
Center unstable	$V^{0+} = V^{cu}$	$|z| \geqslant 1$
$\cup$		
Unstable	$V^+ = V^u$	$|z| > 1$

When $D_a f$ is invertible, V^-, V^{0-}, V^0, V^{0+}, and V^+ for f^{-1} are V^+, V^{0+}, V^0, V^{0-}, and V^-, respectively, for f.

7.3. On the Proof of the Center Stable and Center Unstable Manifold Theorem

Here, again, the graph transform method is used (see Section 6.2), because the study of the center unstable manifold is analogous to the study of the unstable manifold. Note, however, that $D_0\tilde{f}$ is no longer expanding in the E'' direction. Thus, the graph transform of $\psi : E_0''(R) \mapsto E_0'(R)$ is, in general, no longer defined as the graph of a function on $E_0''(R)$. The way out of that difficulty is to consider functions $\psi : E'' \mapsto E_0'(R)$ defined on the whole Banach space E'' and to replace $\tilde{f}$ by a map $\tilde{f}$ equal to $\tilde{f}$ near 0 and suitably close to $D_0\tilde{f}$ on $E'' \times E_0'(R)$. The existence of such an $\hat{f}$ is ensured by the assumption that E'' has the C^r extension property. (Take $\hat{f}(x'', x') = D_0 \tilde{f}(x'', x') + \varphi(x'')(\tilde{f}(x'', x') - D_0 \tilde{f}(x'', x'))$, where φ is like the function in Appendix B.1.) One may then find a fixed point for the graph transformation $\hat{f}_\#$ and prove its smoothness as in the case of the unstable manifold.[27] The locally attracting character of the center unstable manifold corresponds to the characterization (6.2) of the unstable manifold.[28]

7.4. Comparison of the (Un)stable and Center (Un)stable Manifold Theorems

The stable and unstable manifolds of Theorem 6.1 are locally unique, as is clear from the characterizations (6.1) and (6.2). On the other hand, $\mathcal{V}^{0-}$, $\mathcal{V}^{0+}$, and $\mathcal{V}^0$ are not locally unique. (In the discussion of Section 7.3, this is reflected in the nonunique replacement of $\tilde{f}$ by $\hat{f}$.) If f commutes with certain symmetries, it may be possible to respect these symmetries in the construction of the center (un)stable manifold. In spite of the lack of uniqueness, one gets manifolds $\mathcal{V}^{0\pm}$, $\mathcal{V}^0$ invariant under the symmetries. This remark is useful in discussing bifurcations in the presence of a symmetry group.

When $r < \infty$, the (un)stable and the center (un)stable manifolds are all as differentiable as the map f. For a C^∞ or C^ω map f, one may choose a center stable or unstable manifold of class C^r for any finite r, but not of class C^∞ in general.[29]

7.5. Fixed Points of a Flow

Let a be a fixed point for the (local) flow (f^t) and U be a neighborhood of a. Assume that $(x, t) \mapsto f^t x$ is defined and C^r on $U \times \mathbf{R}$, with $1 \leqslant r < \infty$. Then, if the time one map $\tilde{f} = \tilde{f}^1$ satisfies the conditions of Theorem 7.1, one can choose $\mathcal{V}'$, $\mathcal{V}''$ to satisfy the local invariance properties with respect to all f^t. Therefore, if $x \in \mathcal{V}'$, we have $T_x \mathcal{V}' \ni$

---

[27] One has to use a version of the invariant section theorem combining the C^r section theorem and the Hölder section theorem. See Hirsch, Pugh, and Shub [1], Theorem (3.5), Theorem (3.8), and Remark 2 after the latter theorem.

[28] For a proof of Theorem 7.1, see Hirsch, Pugh, and Shub [1], Theorem (5.1), and §5A. The more elementary proofs by Lanford [1], and Marsden and McCracken [1], Section 2, yield a weaker theorem (if f is C^r, φ'' is only C^{r-1}).

[29] For examples of nonuniqueness and nonsmoothness, see A. Kelley [1], Marsden and McCracken [1], (2.6).2; Takens [3].

$X(x) = \frac{d}{dt} f^t x|_{t=0}$, and similarly for $\mathcal{V}''$. Notice, also, that the locally attracting properties hold with the integer n replaced by a real $t \to \infty$.

In order to justify the above statements, we have to show that the proof of Theorem 7.1 can be adapted to the continuous time situation. Let us discuss the case of $\mathcal{V}''$. By analogy with Section 7.3, we want to replace $(\tilde{f}^t)$ by a flow $(\hat{f}^t)$ equal to $(\tilde{f}^t)$ near 0 and suitably close to $D_0 \tilde{f}^t$ in $E'' \times E'_0(R)$. If $(\tilde{f}^t)$ is associated with a $C^{\mathbf{r}}$ vector field X, we define $(\hat{f}^t)$ by the vector field $(x'', x') \mapsto \varphi(x''/R)X(x'', x')$, where φ is like the function in Appendix B.1. We then have

$$\hat{f}^t(x'', x') = \tilde{f}^\tau(x'', x'),$$

where τ is defined by

$$t = \int_0^\tau \varphi((\tilde{f}^\sigma(x'', x'))''/R)^{-1} \, d\sigma \qquad \text{if } \varphi(x'') > 0,$$

and by $\tau = 0$ if $\varphi(x'') = 0$. These formulae define a flow $(\hat{f}^t)$ such that $(x, t) \mapsto \hat{f}^t x$ is $C^{\mathbf{r}}$ even if $(\tilde{f}^t)$ is not associated with a $C^{\mathbf{r}}$ vector field. The construction of $\mathcal{V}''$ by the graph transform method then shows that $\mathcal{V}''$ satisfies the local invariance properties with respect to all f^t.

Unfortunately, the above arguments do not seem to apply to semi-flows. There is thus an interesting open problem: prove a center stable and center unstable manifold theorem for semiflows.[30]

7.6. Periodic Points

One can define a center (un)stable manifold for a periodic point a of a map f by considering a as a fixed point for some iterate f^n. One can also define a center unstable manifold for a periodic orbit of a semiflow by using a Poincaré map.

7.7. Scope of the Graph Transform Method

We have seen how the graph transform method establishes the existence of invariant manifolds through a fixed point. If we do not know that

---

[30]The definition of $(\hat{f}^t)$ given above follows Renz [1]. Marsden and McCracken [1], §2, have noted that it can be used to define center unstable manifolds for flows, but their extension to semiflows is incorrect. Another point of view, and applications, are given in Carr [1].

there is a fixed point, or where it is, the method may still work. For instance, close to the stable and unstable manifolds of a hyperbolic fixed point a for f, the graph transform method will produce a contracting and an expanding manifold for a perturbation $\tilde{f}$, and their intersection will be the new fixed point $\tilde{a}$. Similarly, a center (un)stable manifold will survive perturbation, but a strong stable or unstable manifold for a non-hyperbolic fixed point may not persist (the successive graph transforms walking away to infinity).

The graph transform method extends to the situation where the fixed point a is replaced by orbits (nonperiodic in general) contained in a compact set K. The graphs are now contained in a family of spaces E_x indexed by the points $x \in K$. The assumed compactness of K ensures uniform convergence of the transformed graphs in all E_x. As we shall see, this extension of the method applies to normally hyperbolic manifolds (Section 14) and to hyperbolic sets (Section 15).

8. Attractors, Bifurcations, Genericity

Differentiable dynamical systems exhibit enormous diversity and complexity. Fortunately, there are some general rules of strategy which help in approaching their study, especially when applications to natural phenomena are contemplated.

We first note that, usually, one is less interested in *transients* than in the behavior of natural systems after long times. Therefore, one concentrates attention on *attracting sets*, or *attractors*, for the corresponding dynamical system. A dynamical system may depend on one or more real parameters; such parameters regularly occur in physical applications. The qualitative behavior of the system (e.g., the nature of its attractors) may change when the parameters are varied. Such changes are called *bifurcations*, and their study is a valuable guide in understanding dynamical phenomena. A bifurcation occurs, for instance, when an attracting fixed point a for a map f becomes nonattracting because one of the eigenvalues of $T_a f$ goes through -1 and becomes larger than 1 in absolute value. Note that, in this example, the fixed point is hyperbolic except at the bifurcation. In finite dimension, a fixed point is usually hyperbolic, i.e., hyperbolicity is *generic*. Concentrating attention on generic behavior goes a long way towards putting order in the study of dynamical systems. We shall now review the notions just introduced in somewhat greater detail.

8.1. Attracting Sets

Let the dynamical system (f^t) act on the manifold M.[31] We say that the closed set $\Lambda \subset M$ is an *attracting set* if it has a neighborhood U such that the conditions (a) and (b) or (b′) or (b″) or (b‴) below are satisfied.

(a) *For every neighborhood V of Λ we have $f^t U \subset V$ when t is large enough.*

(b) $f^t \Lambda \supset \Lambda$ *when t is large enough.*

(b′) $f^t \Lambda = \Lambda$ *for all t.*

(b″) $\bigcap_{t \geq T} f^t U = \Lambda$ *for some T.*

(b‴) $\bigcap_{t \geq T} f^t U = \Lambda$ *for all T.*

One can show that, given (a), the conditions (b), (b′), (b″), (b‴) are all equivalent (see Problem 3). Condition (a) implies that $\Lambda \neq \emptyset$ if $U \neq \emptyset$ (otherwise one could take $V = \emptyset$ and obtain $f^t U = \emptyset$). Condition (a) expresses that Λ is attracting, and the conditions (b), (b′), (b″), (b‴) express invariance. We impose here no condition of irreducibility, and thus, the union of two attracting sets is again an attracting set.

A set U such that the above conditions are satisfied is called a *fundamental neighborhood* of the attracting set Λ. We may always choose an open fundamental neighborhood. The open set $W = \bigcup_t (f^t)^{-1} U$ is called the *basin of attraction* of Λ. W consists of those $x \in M$ such that $f^t x \rightarrow \Lambda$ when $t \rightarrow \infty$, and therefore W is independent of the choice of U. If the f^t are invertible (which happens if the time t has negative as well as positive values), we may take Λ equal to the whole manifold M, and then $\Lambda = U = W$.

We have noted in Section 5 that attracting fixed points and periodic orbits were indeed attracting. It is not hard to see that they are attracting sets in the sense of the above definition.

The next result often permits in practice to prove the existence of an attracting set.

8.2. Proposition (on Compact Attracting Sets). *If U is open in M and the closure of $f^t U$ is compact and contained in U for all sufficiently large t, then*

---

[31] As usual, t is discrete or continuous, restricted or not, to be ≥ 0. We impose here no smoothness condition because only continuity of the f^t will be used.

$$\Lambda = \bigcap_{t \geqslant 0} f^t U$$

is a compact attracting set with fundamental neighborhood U.

By assumption, for sufficiently large τ, the closure K of $f^\tau U$ is compact and $K \subset U$. Thus,

$$\bigcap_{t \geqslant 0} f^t K \subset \Lambda \subset \bigcap_{t \geqslant \tau} f^t U \subset \bigcap_{t \geqslant 0} f^t K,$$

and therefore,

$$\Lambda = \bigcap_{t \geqslant 0} f^t K$$

is compact.

Let V be an open neighborhood of Λ. Since $\bigcap_{t \geqslant 0} f^t K \backslash V = \emptyset$, there will, by compactness, be $t_1, t_2, \ldots, t_n$ such that

$$f^{t_1} K \cap \cdots \cap f^{t_n} K \subset V.$$

By assumption, $f^t U \subset U$ for all sufficiently large t, and therefore $f^{t+\tau} U \subset K$ and $f^{t+\tau+\sup t_i} U \subset V$. Property (a) is thus satisfied, and since (b″) holds by definition, Λ is an attracting set.

8.3. ε-Pseudoorbits, Chain-Recurrence, and the Relation ≻

In this section we introduce concepts related to slightly perturbed orbits. Our discussion will lead to a definition of *attractors* different from the attracting sets of Section 8.2.

We assume that $(x, t) \mapsto f^t x$ is continuous and that we have chosen a metric d on M.[32]

A curve (not necessarily continuous), i.e., a family $(x_t)_{t \in [t_0, t_1]}$ with $t_1 \geqslant t_0$, of points of M is called an *ε-pseudoorbit* if

$$d\left(f^\beta x_{t+\alpha}, f^{\alpha+\beta} x_t\right) < \varepsilon$$

whenever $\alpha, \beta \geqslant 0$, $\alpha + \beta \leqslant 1$, and $t, t + \alpha \in [t_0, t_1]$. In the discrete time case, this simply means that $d(f x_t, x_{t+1}) < \varepsilon$ when $t_0 \leqslant t < t_1$.

---

[32]See Proposition B.4.1. Only the *uniform structure* defined by d (see Bourbaki [2]) will be important for the definition of ≻. In particular, d may be replaced by d' if $c_1 d < d' < c_2 d$ with $c_1, c_2 > 0$.

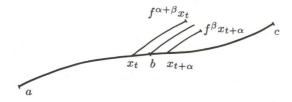

FIG. 13. Putting together two ε-pseudoorbits.

We say that the above ε-pseudoorbit is of length $t_1 - t_0$ and goes from x_{t_0} to x_{t_1}. By putting together two ε-pseudoorbits, with one going from a to b and of length T, the second one going from b to c and of length T', we obtain a 2ε-pseudoorbit (an ε-pseudoorbit in the discrete time case) of length $T + T'$ and going from a to c (see Fig. 13).

We say that the point a is *chain-recurrent* if, for every ε, $L > 0$, there is an ε-pseudoorbit of length $\geq L$ going from a to a. The set of chain-recurrent points is the *chain-recurrent set*.

We now introduce a relation $a \succ b$ that may be read "a goes to b" and that roughly means that there is a slightly perturbed orbit going from a to b. More precisely, for $a, b \in M$, we write $a \succ b$ if, for arbitrarily small $\varepsilon > 0$, there is an ε-pseudoorbit going from a to b.

The relation $\succ$ is a preorder, i.e., it is reflexive ($a \succ a$) and transitive ($a \succ b$ and $b \succ c$ imply $a \succ c$).

The relation $\succ$ is closed, i.e., if $x \succ y$ and $x \to a$, $y \to b$ then $a \succ b$.

The chain-recurrent set is closed.

The proof of these simple properties is left to the reader.

8.4. Basic Classes and Attractors

Write $a \sim b$ if $a \succ b$ and $b \succ a$. Since $\succ$ is a preorder, $\sim$ is an equivalence relation. We denote by $[a]$ the equivalence class of a and write $[a] \geq [b]$ when $a \succ b$. *Every equivalence class is closed* (because the relation $\sim$ is closed).

We say that $[a]$ is a *basic class* if a (and therefore every $x \in [a]$) is chain-recurrent. The chain-recurrent set is thus the union of all basic classes. *An equivalence class $[a]$ is a basic class if and only if a is a fixed point or $[a]$ contains more than one point. An equivalence class $[a]$ is a basic class if and only if $f^t[a] = [a]$ for all t.* (The easy proof of these statements is left to the reader.)

We say that $[a]$ is an *attractor* if it is a minimal equivalence class for the order $\geqslant$. *Every attractor is a basic class* (easy proof).

Notice that, by our definition, the *suitably perturbed* forward orbit of any point x may be expected to come close to some attractor. In fact, one can show that under appropriate conditions, an orbit subjected to small *random* perturbations will *almost certainly* approach an attractor.[33] Random fluctuations are always present in physical experiments and roundoff errors[34] in the computer study of dynamical systems. Therefore, the outputs of many physical or computer experiments are attractors in the above sense. Our general mathematical understanding of attractors is rather poor, and the *Hénon attractor*, for instance, remains quite mysterious. The Hénon attractor[35] appears in the study of the diffeomorphism

$$ f : (x_1, x_2) \mapsto (x_2 + 1 - ax_1^2, bx_1) $$

of $\mathbf{R}^2$ for $a = 1.4$ and $b = 0.3$. This map has a nontrivial attracting set, but it is not easy to see that the attractor visible numerically is not just a long attracting periodic orbit (M. Benedicks and L. Carleson have promised to show that it is not). "Strange" attractors of the Hénon type seem to occur regularly when a diffeomorphism with a homoclinic tangency is perturbed (homoclinic tangencies are defined in Section 16.5).

One frequent feature of attractors is *sensitive dependence on initial condition*. This means that, if x' is close to x, the distance $d(f^t x, f^t x')$, in general, increases exponentially with t (at least as long as the distance remains small). This is a physically important property, and attractors with this feature are called *strange attractors*. The definitions of sensitive dependence on initial condition and strange attractor that we have just given are not mathematically precise. These concepts are nevertheless useful as such in analyzing the results of physical and computer experiments. Time evolutions that show sensitive dependence on initial condition are also called *chaotic*, and the study of *chaos* in natural phenomena has reached considerable popularity and led to a large number of publications.

---

[33] See Ruelle [4].

[34] Roundoff errors are, of course, not truly random, but often behave like random fluctuations for our present purposes.

[35] See Hénon [1], Curry [1], Feit [1], and Choquet [1].

8.5. Nonwandering Points, Invariant Probability Measures

A point $x \in M$ is called *wandering* if it has a neighborhood $\mathcal{N}$ such that, for all sufficiently large t, $\mathcal{N} \cap f^t \mathcal{N} = \emptyset$. The points which are not wandering form the *nonwandering set*. *The nonwandering set is closed and contained in the chain-recurrent set* (easy proof).

If K is a compact subset of M, and ρ a probability measure on K invariant under (f^t), the ergodic theorem (see Appendix C.2) shows that $x \in \operatorname{supp} \rho$ cannot be wandering: *The supports of invariant probability measures are contained in the nonwandering set.*

Fixed points and periodic points are clearly nonwandering. In general, we may say that a point is *recurrent* if it somehow comes back near where it was under time evolution: Nonwandering points and chain-recurrent points are recurrent for two different recurrence notions. Recurrent points play a more important role in the study of differentiable dynamical systems than nonrecurrent points. This is natural because, under time evolution, a nonrecurrent point goes away and never comes back, and there is little one can say about it.

8.6. Bifurcation Theory

Let (f_μ^t) be a dynamical system depending on a parameter μ (the bifurcation parameter). Typically, μ will be a real variable, but it could also be a collection of such variables. If the qualitative nature of the dynamical system changes for a value μ_0 of μ, one says that a bifurcation occurs at μ_0.

The bifurcation idea may be formalized by introducing the concept of *structural stability*. Let $\mathcal{D}$ be a specific space of differentiable dynamical systems (for instance, the space of C^2 diffeomorphisms on a compact manifold M with the usual topology, or a space of semiflows). A point (f^t) of $\mathcal{D}$ is said to be structurally stable if, for every (g^t) sufficiently near to (f^t) in $\mathcal{D}$, there is a homeomorphism h of M that maps (f^t)-orbits to (g^t)-orbits, preserving the order of the points on the orbits. In the discrete time case, this means that the map g^1 is topologically conjugate to f^1 (as in the Grobman–Hartman Theorem 5.2). In the continuous-time case, we allow a reparametrization of the orbits (in particular, a closed orbit for (g^t) need not have the same period as the corresponding closed orbit for (f^t)). Let Σ be the set of points in $\mathcal{D}$ that are not structurally stable; Σ is a closed set called the *bifurcation set*. If we think now of (f_μ^t) as a curve in $\mathcal{D}$, the bifurcations correspond to the intersections of that curve with Σ. Indeed, the qualitative nature of the dynamical system can change only when the bifurcation set Σ is crossed. If a piece Σ_1 of Σ is a submanifold of codimension 1 of $\mathcal{D}$, we say that it corresponds to a *codimension 1 bifurcation. Codimension k bifurcations* are defined similarly, and will normally be seen when the bifurcation parameter μ consists of at least k real variables.

If Σ would be the union of a nice collection of manifolds of various codimensions in a background of structurally stable dynamical systems, we would have a very satisfactory picture of dynamical systems and their bifurcations. Things, however, are not so simple: There exist nonempty open sets of $\mathcal{D}$ consisting of nonstructurally stable dynamical systems. The concept of structural stability may be weakened to Ω-*stability*, which is structural stability restricted to the nonwandering set Ω. Again, however, $\mathcal{D}$ contains nonempty open sets of non-Ω-stable dynamical systems.[36] Structural stability and Ω-stability are thus of more limited significance than anticipated. Their study, promoted, among others, by R. Thom and S. Smale,[37] has, however, been important for our understanding of dynamical systems.

For the purposes of Part 2, we shall limit our ambition to investigating specific bifurcation phenomena that are important in applications. The concepts of structural stability and Ω-stability will be useful in this study, but weaker notions will also occur.

The bifurcation theory of attracting sets and attractors is particularly interesting. We discuss here a simple phenomenon. Suppose that the system (f_μ^t) has an attractor A_μ that changes continuously with the real parameter μ. If $x_0 \in A_{\mu_0}$, and if μ is increased *adiabatically*, i.e., very slowly with t, we may expect that $f_\mu^t x_0$ will remain very close to A_μ. If the attractor A_μ disappears for some value μ' of μ, $f_\mu^t x_0$ may jump to another attractor B_μ. If μ is now decreased, $f_\mu^t x_0$ will not jump back to A_μ, but will stay near B_μ (see Fig. 14). This nonreversible behavior

---

[36]Non-Ω-stability and, therefore, nonstructural stability occur, for instance, if $\mathcal{D}$ consists of diffeomorphisms of a manifold of dimension $m \geqslant 2$. See Section 16.7. For further results on structural stability and Ω-stability, see Appendix D.17, see also Problem 7 of Part 2.

[37]See in particular Thom [1] and Smale [3].

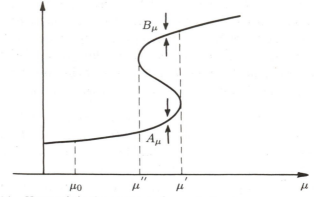

FIG. 14.   Hysteresis in the presence of several attractors.

is called *hysteresis*.[38] It is one expression of the fact that the choice of attractor made by a system depends on the past history of the system.

8.7. Genericity

We have just seen that the set $\mathcal{D}\backslash\Sigma$ of structurally stable dynamical systems could not, in general, be considered a large subset of $\mathcal{D}$. On the other hand, a dense open subset of $\mathcal{D}$ could be viewed as large. A weaker but useful notion is that of a residual set. We say that $\mathcal{G}$ is a *residual* subset of $\mathcal{D}$ if it contains a countable intersection of dense open subsets of $\mathcal{D}$. If we assume that the topology of $\mathcal{D}$ is compatible with a metric for which $\mathcal{D}$ is complete, we obtain that $\mathcal{G}$ is dense in $\mathcal{D}$ as a consequence of Baire's theorem.[39] Notice that a countable intersection of residual sets is again residual. If a property of dynamical systems is true in a residual subset of $\mathcal{D}$, this property is said to be *generic*.

It is good to know whether a property is generic or not, but it should be understood that *generic* does not imply *usually true*. One could say that a property is usually true if it were true on a set $\mathcal{U} \subset \mathcal{D}$ such that $\mathcal{D}\backslash\mathcal{U}$ has measure 0 for some natural measure class. *There is, however, no natural measure class on spaces of differentiable dynamical systems.*

---

[38] If a piece of magnetic material is submitted to a magnetic field, the magnetization will depend on the history of variations of the field, and not just on its final value. This is the usual phenomenon of (magnetic) hysteresis. A model of magnetic hysteresis can be given in terms of bifurcations of a dynamical system (see Erber, Guralnik, and Latal [1]).

[39] See Appendix A.3.

One is thus led to using concepts like genericity, but with due caution.[40] For instance, if we know that a property is true except on a differentiable submanifold of $\mathcal{D}$, it is most reasonable to say that the property is usually true, provided $\mathcal{D}$ is the appropriate space for the problem under consideration.[41] Incidentally, let us remark that one often says that the property "P" is *nongeneric* to mean that "non P" is generic (this is definitely an abuse of language). Also, the word *generic* is sometimes used in a loose sense rather than with the precise meaning defined above.

8.8. *Generic Properties of Diffeomorphisms and Flows on a Compact Manifold*

Below we give two famous examples, one due to Kupka–Smale and one to Pugh, of generic properties of diffeomorphisms and flows on a compact manifold M. However, first we must show how a complete metric compatible with the topology can be put on the space $\text{Diff}^{\mathbf{r}}(M)$ of $C^{\mathbf{r}}$ diffeomorphisms of M, or the space $\mathcal{X}^{\mathbf{r}}(M)$ of $C^{\mathbf{r}}$ vector fields on M, when $\mathbf{r} \leqslant \infty$. The idea is to describe $f \in \text{Diff}^{\mathbf{r}}$, or $x \in \mathcal{X}^{\mathbf{r}}$, in terms of a finite family of maps from open sets in $\mathbf{R}^m$ to $\mathbf{R}^n$, and to use the standard $C^{\mathbf{r}}$ norms for these maps.

By Whitney's theorem, we may identify M with a C^{∞} submanifold of $\mathbf{R}^{2m+1}$. Also let $(U_\alpha, \varphi_\alpha, \mathbf{R}^m)$ be a finite collection of charts covering M.[42] To $f \in \text{Diff}^{\mathbf{r}}(M)$ we associate the maps $f_\alpha = f\varphi_\alpha^{-1}$ from $\varphi_\alpha U_\alpha$ to $\mathbf{R}^{2m+1}$. Using the definition of $\| \cdot \|_{U,\mathbf{r}}$ in Appendix B.1, we write

$$\|f\|_{\mathbf{r}} = \sum_\alpha \|f_\alpha\|_{\varphi_\alpha U_\alpha, \mathbf{r}}$$

for finite $\mathbf{r}$, and define the desired metric on $\text{Diff}^{\mathbf{r}}(M)$ by

$$d_{\mathbf{r}}(f, g) = \|f - g\|_{\mathbf{r}} + \|f^{-1} - g^1\|_{\mathbf{r}}.$$

For $\text{Diff}^{\infty}(M)$, we may write

$$d_\infty(f, g) = \sum_{r=1}^{\infty} 2^{-r} \frac{d_r(f, g)}{1 + d_r(f, g)}.$$

---

[40] Notice that a dense open set in the interval $[0, 1]$ may have arbitrarily small Lebesgue measure. Take an open interval of length $\varepsilon/n!$ around the nth rational in $[0, 1]$ (for some enumeration of the rationals), the union of these intervals has length $< e.\varepsilon$. The intersection of the sets just constructed for a sequence of values of ε tending to zero is residual of measure 0.

[41] See the remarks in Sections 2.5 and 2.6.

[42] We assume that $\dim M = m$: It would be easy to treat the case where M has several components of different dimensions.

To a vector field $X \in \mathcal{X}^{\mathbf{r}}(M)$ are associated functions $X_\alpha : \varphi_\alpha U_\alpha \mapsto$ $\mathbf{R}^m$ (see Section 1.3), and we may define

$$d_{\mathbf{r}}(X,Y) = \sum_\alpha \|X_\alpha - Y_\alpha\|_{\varphi_\alpha U_\alpha, \mathbf{r}}$$

for **r** finite, and

$$d_\infty(X,Y) = \sum_{r=1}^\infty 2^{-r} \frac{d_r(X,Y)}{1 + d_r(X,Y)}.$$

It is easy to see that these metrics are complete and define the $C^{\mathbf{r}}$ topologies on $\mathrm{Diff}^{\mathbf{r}}$ and $\mathcal{X}^{\mathbf{r}}$.

If P is a hyperbolic periodic orbit for a diffeomorphism or flow (or a fixed point for a flow), we have defined the *local* stable and unstable manifolds V_P^+ and V_P^- in Section 6.3. *Global* stable and unstable manifolds are defined by

$$\mathcal{W}_P^- = \bigcup_{T \geqslant 0} f^{-T} V_P^-,$$

$$\mathcal{W}_P^+ = \bigcup_{T \geqslant 0} f^T V_P^+.$$

These sets are manifolds locally[43] and this is sufficient to define transversality.

8.9. Theorem (Kupka–Smale).[44] *Let $1 \leqslant \mathbf{r} \leqslant \infty$ and let $\mathcal{G} \subset \mathrm{Diff}^{\mathbf{r}}(M)$ consist of those diffeomorphisms for which all periodic orbits are hyperbolic, and for which the intersection of the global unstable manifold of each periodic orbit with the global stable manifold of each periodic orbit is transversal. Then $\mathcal{G}$ is residual in $\mathrm{Diff}^{\mathbf{r}}(M)$. Similarly for $\mathcal{X}^{\mathbf{r}}(M)$.*

Consider the diffeomorphisms $f \in \mathrm{Diff}^{\mathbf{r}}$ such that the periodic orbits are hyperbolic when the period is $\leqslant T_0$, and such that the corresponding *local* stable and unstable manifolds $f^{-T_1} V_P^-$, $f^{T_1} V_Q^+$ intersect transversally. One shows that the set G of such diffeomorphisms is open and dense. By letting T_0, T_1 go to infinity, one obtains a residual set $\mathcal{G}$. Similarly for flows.

---

[43]$\mathcal{W}_P^+$ and $\mathcal{W}_P^-$ may be dense in M; see Problem 2.

[44]This theorem was proven independently by Kupka [1] and Smale [1]. A very accessible proof is given by Palis and de Melo [1], Chapter 3. An extension to maps is given by Shub [1].

8.10. Theorem (Pugh).[45] *Let $\mathcal{G}$ be the subset of* $\text{Diff}^1(M)$ *or* $\mathcal{X}^1(M)$ *consisting of diffeomorphisms or flows such that the periodic orbits are dense in the nonwandering set Ω (i.e., the closure of the union of the periodic orbits is Ω). Then $\mathcal{G}$ is residual in* Diff^1 *or* $\mathcal{X}^1$.

This is a form of Pugh's *closing lemma*. It is a difficult result, which has been proven only for the C^1 topology.

Note

Our general presentation of differentiable dynamical systems allows the time to be discrete (maps and diffeomorphisms) or continuous (semiflows and flows). We consider Banach manifolds of finite or infinite dimension, and we discuss $C^{(r,\alpha)}$ differentiability (α-Hölder rth derivative). This degree of generality is desirable for applications and does not require much extra work.

The discussion of manifolds in Sections 1–3 is made more precise in Appendix B, which contains the appropriate references. Sections 4–7 deal with fixed points and periodic orbits: the Poincaré map, hyperbolicity, the Grobman–Hartman theorem, and invariant manifolds. Further details and other viewpoints (for this and Part 2) may be found in the monographs by Moser [1], Shub [2], Palis and de Melo [1], Marsden and McCracken [1], Guckenheimer and Holmes [1], and the excellent lecture notes by Newhouse [3] and Lanford [3]. See also Iooss [1], Nitecki [1], Abraham and Marsden [1], and the pictorial presentation of Abraham and Shaw [1]. A basic source on invariant manifolds is Hirsch, Pugh, and Shub [1]. In Section 8, the definition of attractors and attracting sets follows Ruelle [4], based on ideas of Conley [1].

The theory of differentiable dynamical systems may be said to begin with Poincaré (in particular, the famous *Nouvelles Méthodes de la Mécanique Céleste*, see Poincaré [3]). The modern theory has been developed particularly along two different lines. One direction of research is centered on the Kolmogorov–Arnold–Moser theorem and small denominator problems and has applications mainly to Hamiltonian systems. The other direction of research deals with hyperbolic systems and is more useful for dissipative systems. In its foundations, the study of hyperbolic systems owes a lot to the Russian school (see, in particular, Anosov [1]), and to Thom and Smale. Smale's paper [3] is a

---

[45] See Pugh [1], Pugh and Robinson [1].

masterpiece of mathematical literature. Thom has presented his mathematical ideas together with a controversial program of applications to natural phenomena in a provocative book (Thom [1]), which has exerted a very significant influence. The study of differentiable dynamics is now actively pursued in America—North and South—and Europe—East and West—and is developing elsewhere.

At this point, the reader is advised to go rapidly through Appendices A, B, and C if he or she has not already done so earlier.

Problems

1. Hyperbolicity. Let E_1, E_2 be separable Hilbert spaces, with origins O_1, O_2. Let $f_1 : E_1 \mapsto E_2$ be a linear isometry on a proper subspace of E_2, and $f_2 : E_2 \mapsto E_1$ a linear map such that $f_2 f_1$ is equal to multiplication by $\alpha > 0$ on E_1, and f_2 vanishes on the orthogonal complement of $f_1 E_1$. We define the manifold M to be the disjoint union of E_1 and E_2, and the map $f : M \mapsto M$ to have restriction f_1, f_2 to E_1, E_2. Show that O_1 is a fixed point for f^2, which is attracting if $\alpha < 1$, repelling if $\alpha > 1$; $\{O_1, O_2\}$ is a periodic orbit of period 2 of f, which is hyperbolic if $\alpha \neq 1$, attracting if $\alpha < 1$, but not repelling if $\alpha > 1$.

[Use the definitions in Section 5.5 and study the spectra of $f_2 f_1$ and $f_1 f_2$].

2. (Global stable and unstable manifolds). Let a be a hyperbolic fixed point for a $C^{\mathbf{r}}$ map $f : M \mapsto M (\mathbf{r} \geqslant 1)$, and let V_a^-, V_a^+ be local stable and unstable manifolds. We define the global stable and unstable manifolds by

$$W_a^- = \left\{ x \in M : \lim_{n \to \infty} f^n x = a \right\},$$

$$W_a^+ = \left\{ x_0 : \exists (x_n)_{n \geqslant 0} \text{ with } f x_{n+1} = x_n \text{ and } \lim_{n \to \infty} x_n = a \right\}.$$

(a) Show that

$$W_a^- = \bigcup_{n \geqslant 0} f^{-n} V_a^-, \quad W_a^+ = \bigcup_{n \geqslant 0} f^n V_a^+.$$

If f is a diffeomorphism, each $f^{\pm n} V_a^{\pm}$ is $C^{\mathbf{r}}$. If $Tf : TM \mapsto TM$ is surjective and V_a^- has finite codimension (or M is a Hilbert manifold), then $f^{-n}V_a^-$ is C^r. If $Tf : TM \mapsto TM$ is injective and V_a^+ finite dimensional, then $f^n V_a^+$ is C^r. The sets $W_a^\pm$ are thus locally smooth manifolds.

[Use the implicit function theorem; the finite (co)dimension or Hilbert conditions are only to obtain closed linear spaces with closed complements.]

Note: If f is a diffeomorphism of a finite dimensional manifold, one can show that $W_a^\pm$ is the image of $V_a^\pm$ by an injective C^r immersion tangent to the identity at a.

(b) Let $h : \mathbf{R}^2 \mapsto \mathbf{R}^2$ be the map defined by the matrix $\begin{pmatrix} 1 & 1 \\ 1 & 2 \end{pmatrix}$ and $f : T^2 \mapsto T^2$ the induced diffeomorphism on the torus $T^2 = \mathbf{R}^2/\mathbf{Z}^2$. Compute $W_a^\pm$ and show that they are dense in T^2.

3. (Attracting sets). Throughout what follows, we assume that condition (a) of Section 8.1 is satisfied. Prove

$$(1) \qquad \Lambda \supset \bigcap_{t \geq T} f^t U \qquad \text{for every } T.$$

If $x \notin f^\tau \Lambda$, deduce that $f^t U \subset (f^\tau)^{-1}(M\backslash\{x\})$, hence $f^t U \subset M\backslash\{x\}$ for large t, hence

$$(2) \qquad \bigcap_{t \geq T} f^t U \subset f^\tau \Lambda \qquad \text{for large } T.$$

Using (2), show that $(b'') \Rightarrow (b)$. From (1) show that (b) implies

$$(3) \qquad \Lambda = \bigcap_{t \geq T} f^t U \qquad \text{for large } T,$$

and therefore

$$(4) \qquad f^\tau \Lambda \subset \Lambda \qquad \text{for all } \tau.$$

From (4), deduce that $(b) \Rightarrow (b')$. From (1) and (3), deduce that $(b) \Rightarrow (b''')$.

Conclude that properties (b), (b'), (b''), (b''') of Section 8.1 are all equivalent.

4. (Stable and unstable sets of general points). Let (f^t) be a dynamical system on M. For any $a \in M$, we write

$$W_a^- = \{x \in M : \lim_{t \to \infty} \text{dist}(f^t x, \ f^t a) = 0\},$$

and if f is bijective,

$$W_a^+ = \{x \in M : \lim_{t \to \infty} \text{dist}(f^{-t} x, \ f^{-t} a) = 0\}.$$

We call W_a^- the *stable set* of a, and W_a^+ the *unstable set* of a. Show that, if a is chain recurrent, $W_a^- \succ a \succ W_a^+$. Thus, if a, b are chain recurrent and $W_a^+ \cap W_b^- \neq \emptyset$, then $a \succ b$. If $a \sim b$, then $W_a^+ \cap W_b^- \subset [a]$. If a is an attractor, then $W_a^+ \subset [a]$.

5. (Inclination lemma or λ-lemma).

Let V_a^-, V_a^+ be the stable and unstable manifolds of the hyperbolic fixed point a for a sufficiently smooth map f. If Σ is a manifold intersecting V_a^+ transversally at a single point, the λ-lemma[46] asserts that, when $n \to \infty$, $f^{-n}\Sigma$ (suitably restricted) tends to the stable manifold V_a^-. (Similarly with f, $V^\pm$ replaced by f^{-1}, $V^\mp$.) Furthermore, if $\dim V_a^+ = 1$ and A^+ is the expanding eigenvalue of $D_a f$, the manifolds $f^{-n}\Sigma$ approach V_a^- like $|A^+|^{-n}$.[47] A more precise and purely local reformulation of these results is given below.

(a) Let $E^\pm$ be Banach spaces, and $f : E_0^+(R) \oplus E_0^-(R) \mapsto E^+ \oplus E^-$ be $C^{(1,\alpha)}$ (C^1 is enough if E^- has finite dimension). We let

$$f(x, y) = (A^+ x + F(x, y), A^- y + G(x, y)),$$

where $A^\pm$ are linear operators and

$$\|A^-\| = a < 1, \qquad \|(A^+)^{-1}\| = b^{-1} < 1,$$

$$D_{(0,0)} F = 0, \qquad D_{(0,0)} G = 0.$$

We also assume, as we may, that

$$F(0, y) = 0, \qquad G(x, 0) = 0.$$

---

[46] See Palis [2].
[47] See Palis [5].

Show that if $\psi : E_0^-(\delta) \mapsto E_0^+(R)$ is C^1 for some $\delta > 0$, the graph transforms $f_{\#}^{-n}\psi$ (defined on $E_0^-(R)$ for large n) tends to 0 in the C^1 topology when $n \to \infty$.[48]

[The graph transform is defined in Section 6.2. Since $f^{-n}(\psi(0), 0) \to (0,0)$ when $n \to \infty$, the box $E_0^+(R) + E_0^-(R)$ may be replaced by a smaller box $E_0^+(R^{**}) \oplus E_0^-(R^*)$. Indeed, if we can show that $f_{\#}^{-n}\psi$ is C^1 small on $E_0^-(R^*)$, $f_{\#}^{-n-n'}\psi$ is C^1 small on $E_0^-(R)$ for suitably chosen n'. Fix R^* such that, in $E_0^+(R^*) \oplus E_0^-(R^*)$, the derivatives DF, DG have norm less than k, with

(1) $\qquad a + k < 1, \quad b - k > 1, \quad k < \dfrac{(b - k - 1)^2}{4}.$

The choice of R^{**} will depend on some $\varepsilon > 0$ to be fixed later. By assumption, the derivative $D^2 F$ of $F(x, y)$ with respect to the second argument is *uniformly continuous*, and vanishes for $y = 0$. Therefore, we may take $R^{**} \in (0, R^*]$ such that

(2) $\qquad \|D_{(x,y)}^2 F\| \leqslant \varepsilon \qquad$ for $(x, y) \in E_0^+(R^{**}) \oplus E_0^-(R^*).$

Let $Z_n = ((f_{\#}^{-n}\psi)(0), 0)$, $(u, v) = (D_{Z_n} f^n)(u_n, v_n)$ and $\lambda_n = \|u_n\|/\|v_n\|$. (The *inclination* λ gives its name to the λ-lemma.) Using (1), show that

$$\lambda_n \leqslant \frac{\lambda_0}{(b - k)^n} + k \sum_1^n \frac{1}{(b - k)^j}$$

so that, for large n, $\lambda_n \leqslant (b - k - 1)/4$. Restricting $f_{\#}^{-n}\psi$, if necessary, to $E_0^-(\delta^*)$ with small δ^*, we may assume that the tangent vectors to the graph of $f_{\#}^{-n}\psi$, for large n, have inclination $\lambda_n^* \leqslant (b - k - 1)/2$. Use (2) to study the inclination of the tangent vectors to $f_{\#}^{-n}\psi$, and prove that $f_{\#}^{-n}\psi \to 0$ in C^1.]

(b) Similarly to (a), if $\widetilde{\psi} : E_0^+(\delta) \mapsto E_0^-(R)$ is C^1, the graph transforms $f_{\#}^n \widetilde{\psi}$ tend to 0 in C^1 when $n \to \infty$.

(c) Returning to the assumptions of (a), let f be of class $C^{(1,1)}$ and $\dim E^+ = 1$. Suppose that the curve $\zeta : \mathbf{R} \mapsto E_0^+(R) \oplus E_0^-(R)$

---

[48]This result can be extended to $C^{\mathbf{r}}$ for $\mathbf{r} > 1$ (one goes from the $C^{\mathbf{r}}$ case to the $C^{\mathbf{r}-1}$ case by the trick of considering the map $(x, u) \mapsto (fx, D_x fu)$ instead of f).

is C^1 and transversal to the stable manifold $E_0^-(R)$ at λ_∞. Then, for sufficiently large n, there are λ_n such that $f^n\zeta(\lambda_n) \in$ graph ψ. Furthermore, $\lim_{n\to\infty} (\lambda_n - \lambda_\infty)/(\lambda_{n+1} - \lambda_\infty) = A^+$ (where A^+ is the expanding eigenvalue of $D_{(0,0)}f$).

[By assumption, we may take here $\|D^2 F(x,y)\| \leqslant K\|x\|$. From this, deduce by induction that the inclination λ of a tangent vector to $f_\#^{-n}\psi$ at (x,y) for large n satisfies $\lambda \leqslant C\|x\|$ for some constant C. Now, by taking $\dim E^+ = 1$, we may compare the points (ξ_n, η_n) of the curve ζ such that $f^n(\xi_n, \eta_n) \in \Sigma$, and the points $(x_n, 0)$ of E^+ with the same property. If ℓ is an upper bound to the distance of ζ and E^+, we find $e^{-C\ell} \leqslant |\xi_n/x_n| \leqslant e^{C\ell}$, hence

$$e^{-2C\ell}|x_n/x_{n+1}| \leqslant |\xi_n/\xi_{n+1}| \leqslant e^{2C\ell}|x_n/x_{n+1}|.$$

Since we may replace the curve ζ by a curve $f^m\zeta$ closer to E^+, the conclusion follows.]

6. (Grobman–Hartman theorem).

(a) Let 0 be an attracting fixed point for the local flow (f^t) defined by a C^1 vector field X on a neighborhood of 0 in $\mathbf{R}^m$. Let B be the open ball limited by a sphere S of sufficiently small radius centered at 0. Show that there is a unique homeomorphism $h : B \mapsto B$ such that

$$h(e^{t(D_0 X)}x) = f^t x$$

for $x \in S$ and $t > 0$. (Note that h is continuous but need not be differentiable at 0.) Verify that this gives a proof of the Grobman–Hartman theorem for a flow at an attracting or repelling fixed point in finite dimension.

(b) Similarly prove the Grobman–Hartman theorem for a diffeomorphism f at an attracting or repelling fixed point in finite dimension.

[Let O be attracting for f, and let B, S be as in (a). Define a homeomorphism h of $\overline{B}\backslash(D_0 f)B$ to $\overline{B}\backslash fB$ such that $hx = x$ and $h(D_0 f)x = fx$ if $x \in S$. Verify that h extends uniquely to a homeomorphism $B \mapsto B$ such that $h \circ D_0 F = f \circ h$.]

(c) Let a be a hyperbolic fixed point for a flow or a diffeomorphism of a finite-dimensional manifold M. Use Problem 5 to construct continuous invariant foliations $(\mathcal{F}_\xi^-)$, $(\mathcal{F}_\eta^+)$ of a neighborhood of a such that $\mathcal{F}_0^- = V_a^-$, $\mathcal{F}_0^+ = V_a^+$. (The *leaves* $\mathcal{F}_\xi^-$ are thus disjoint manifolds covering a neighborhood $\mathcal{O}$ of a, depending continuously on ξ, such that $f^t \mathcal{F}_\xi^- | \mathcal{O} \subset \mathcal{F}_{f^t \xi}^-$ and that $\mathcal{F}_0^-$ reduces to the stable manifold of a, similarly for $(\mathcal{F}_\eta^+)$.) Show that, by a suitable choice of the coordinates ξ, η, one obtains a proof of the Grobman–Hartman theorem.[49]

[Use part (a) or (b).]

7. (Diffeomorphisms that are not time one maps).   Let M be any compact manifold with $\dim M \geqslant 1$. Show that, in every C^∞ neighborhood of the identity map, there is a diffeomorphism f that does not embed in a flow, i.e., f is not the time one map of a flow.

[Choose f having a hyperbolic periodic orbit of period > 1.]

[49]This proof was pointed out to me by Jacob Palis.

2   Bifurcations

Est opus egregium sacros iam scribere libros
nec mercede sua scriptor et ipse caret.

— *Alcuinus*

In this chapter we discuss a number of specific bifurcation phenomena. The most important bifurcations are those of codimension 1. The best-understood bifurcations are those for which a local analysis is possible. These two facts will dominate our study.

9. Bifurcations of Fixed Points of a Map

Let U be an open subset of the Banach space E, J an interval of $\mathbf{R}$, and $f_\mu : U \mapsto E$ a map defined for $\mu \in J$. We have thus a *parametrized family* of maps (f_μ). We assume that $(\mu, x) \mapsto f_\mu(x)$ is $C^{\mathbf{r}} : J \times U \mapsto E$ with $\mathbf{r} \geqslant 1$. This setting is appropriate for the local study of codimension 1 bifurcations of maps.

Suppose that $\mu_0 \in J$ and that f_{μ_0} has a fixed point x_0 such that $D_{x_0} f_{\mu_0} - 1$ is invertible. Then, the Implicit Function Theorem B.3.3 implies that x_0 is an isolated fixed point of f_{μ_0} and that there is a $C^{\mathbf{r}}$ function (locally unique)$\mu \mapsto x(\mu)$ from a neighborhood of $\mu_0$in J to

U such that $x(\mu)$ is an isolated fixed point of f_μ and $x(\mu_0) = x_0$. This situation occurs, in particular, if x_0 is a hyperbolic fixed point of f_{μ_0}, and $x(\mu)$ may then also be assumed to be hyperbolic by continuity. The stable and unstable manifolds of $x(\mu)$ vary continuously with μ (see Theorem 6.1). If $D_{x_0} f_{\mu_0}$ is hyperbolic and invertible,[1] f_μ is locally topologically conjugate to $D_{x_0} f_{\mu_0}$ for μ close to μ_0 (see Theorem 5.2). In view of these remarks, we should look for bifurcations at values of μ such that $x(\mu)$ is not hyperbolic.

9.1. Considerations of Genericity

It is good to realize right away that *curves of fixed points of f_μ in $J \times E$, generically do not cross.* The *pitchfork bifurcation* of Figure 15 is thus nongeneric. To see this, we take for simplicity $E = \mathbf{R}^m$, and claim that

$$\mathcal{F} = \{(\mu, x) \in J \times U : f_\mu x = x\}$$

generically is a *nonsingular* curve. In other words, at each $(\mu, x) \in \mathcal{F}$, there is, up to a multiplicative factor, only one tangent vector, i.e., a vector (μ', x') satisfying

(9.1) $$\frac{\partial f_\mu}{\partial \mu} \cdot \mu' + (D_x f_\mu - 1)x' = 0.$$

We make a naive verification of this fact by counting the number of conditions necessary to create a singularity of $\mathcal{F}$: For $m + 1$ variables (μ and X), there are $m + 2$ conditions (m for $f_\mu x = x$, and 2 to ensure that the $(m+1) \times m$ matrix $(\partial_\mu f, D_x f)$ of (9.1) has rank $m - 1$). Thus $\mathcal{F}$ is in general not singular (it has no self-intersection, isolated point, etc.).[2] The above $m + 2$ conditions mean that $(\mu, x) \mapsto f_\mu x - x$ is not transversal to $\{0\} \subset \mathbf{R}^m$, therefore *nonsingularity* is the same thing as *smoothness*, as should naturally be the case. The generic smoothness of $\mathcal{F}$ can be proved rigorously by use of *transversality theory*.[3]

Altogether, instead of Fig. 15, we should think of Fig. 16 as giving a generic picture of fixed points of a map depending on a parameter.

---

[1] We do not discuss here the bifurcation that corresponds to the vanishing of an eigenvalue of $D_{x_0} f_{\mu_0}$. For a map in one dimension, this expresses the *superstability* of the fixed point x_0.

[2] If E is infinite dimensional, one can argue similarly, provided one has sufficient control of the spectrum of $D_x f_\mu$ (a suitable condition would be that $D_x f_\mu$ is the sum of a compact operator and a contraction).

[3] See Problem 1.

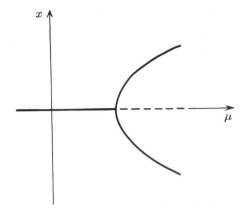

FIG. 15. A nongeneric bifurcation (pitchfork): an attracting fixed point loses its attracting character while two new attracting fixed points are created.

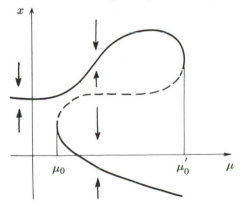

FIG. 16. A generic bifurcation diagram with a saddle-node bifurcation at $\mu = \mu_0$ (case A) and an inverted saddle-node bifurcation at μ_0' (case B).

As we have seen in Section 2.5, there may be a group G of symmetries acting on M, such that f_μ is constrained to be G-equivariant. *The presence of a symmetry group changes the generic bifurcations.* For instance, in the presence of the symmetry $x \mapsto -x$ in $\mathbf{R}^m$, the pitchfork bifurcation of Fig. 15 becomes generic. In situations relevant to physics, M is often a linear space, and G acts linearly on M. The discussion of generic bifurcations in this setup is useful for applications, in particular to hydrodynamics.[4]

[4]See Ruelle [1], Rand [1], Golubitsky and Steward [1], Chossat and Golubitsky [1], and also Sattinger [1].

Note that bifurcations of Hamiltonian (conservative) dynamical systems (Section 2.6) are special and will not be considered here. The generic bifurcations discussed below are relevant to the analysis of dissipative physical systems.

9.2. Saddle-Node, Flip, and Hopf Bifurcations

Suppose now that $(\mu_0, x_0) \in \mathcal{F}$, and that x_0 is not hyperbolic. This means that part of the spectrum of $D_{x_0} f_{\mu_0}$ is on the unit circle, and there are generically three ways in which this may occur.[5]

(a) The only eigenvalue of $D_{x_0} f_{\mu_0}$ on the unit circle is simple and equal to 1. One says that x_0 is a *saddle-node*. Since $D_{x_0} f_{\mu_0} - 1$ is not invertible, we cannot apply the implicit function theorem to get a curve $\mu \mapsto x(\mu)$. We shall see in Section 11 that, generically, a saddle node at $\mu = \mu_0$ corresponds to the coalescence of two hyperbolic fixed points present either for $\mu > \mu_0$ or $\mu < \mu_0$. These two forms of *saddle node bifurcation* are illustrated in Fig. 16.

(b) The only eigenvalue of $D_{x_0} f_{\mu_0}$ on the unit circle is simple and equal to -1. The implicit function theorem gives a curve $\mu \mapsto x(\mu)$ through (μ_0, x_0). We shall see in Section 12 that, generically, a *flip bifurcation* occurs, where a periodic orbit of period 2 is present either for $\mu > \mu_0$ or for $\mu < \mu_0$. These two cases are illustrated in Figs. 17A and B. The flip bifurcation is also called *period doubling* or *subharmonic* bifurcation (these names are used especially for the corresponding bifurcation of a periodic orbit, Section 10.1).

(c) There are two simple complex conjugate eigenvalues $\alpha, \bar{\alpha}$ of $D_{x_0} f_{\mu_0}$ on the unit circle. The implicit function theorem gives a curve $\mu \mapsto x(\mu)$ through (μ_0, x_0). We shall see in Section 13 that, generically, a *Hopf bifurcation* occurs, where a curve γ_μ (circle) invariant under f_μ is present either for $\mu > \mu_0$ or for $\mu < \mu_0$. These two cases are illustrated in Figs. 18A and B.

Figure 19 shows a Hopf bifurcation for a 2-dimensional flow. Near the fixed points, the orbits spiral inward for $\mu < \mu_0$ and outward for $\mu > \mu_0$ as dictated by *linear theory*. (Linearization of f_μ^t near the fixed point

---

[5]We assume that either E is finite dimensional, or else that one has suitable control over the spectrum of $D_x f_\mu$ (see footnote 2).

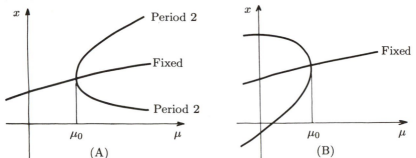

FIG. 17A and B. (A) Flip bifurcation: a family of periodic orbits of period 2 is created when μ increases beyond μ_0. (B) Inverted flip bifurcation: a family of periodic orbits of period 2 disappears when μ increases beyond μ_0. Note that a flip bifurcation for f_μ corresponds to a (nongeneric) pitchfork bifurcation for f_μ^2.

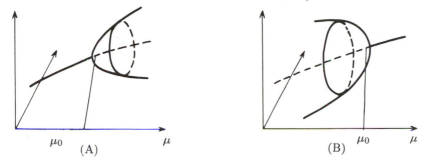

FIG. 18A and B. (A) Hopf bifurcation: a family of invariant circles appears when μ increases beyond μ_0. (B) Inverted Hopf bifurcation: a family of invariant circles disappears when μ increases beyond μ_0.

may be justified by the Grobman–Hartman Theorem 5.4.) The Hopf bifurcation is a nonlinear effect: For $\mu > \mu_0$ the orbits still spiral inward at some distance from the fixed point, whereas they spiral outward in the immediate vicinity. A closed orbit is then necessarily trapped between the inward and outward spiraling orbits.

9.3. Use of a Center Manifold

Suppose that x_0 is a fixed point of f_{μ_0}, and that a bifurcation occurs at (μ_0, x_0). If the dimension of $V_{x_0}^{0+}$ is finite, one can reduce the study of f_μ to that of a map $\tilde{f}_\mu$ in finite dimension. To see this, assume that $(\mu, x) \mapsto f_\mu x$ is C^r near (μ_0, x_0), with $1 \leqslant r < \infty$. A map $F : J \times U$ to $\mathbf{R} \times M$ is defined by $F(\mu, x) = (\mu, f_\mu x)$, and F has (μ_0, x_0) as a fixed point. Notice that $D_{(\mu_0, x_0)}F$ has the same spectrum as $D_{x_0}f_{\mu_0}$,

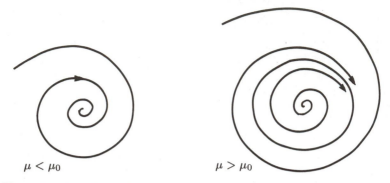

$\mu < \mu_0$ $\mu > \mu_0$

FIG. 19.   Hopf bifurcation for a flow in two dimensions.

plus an eigenvalue 1 corresponding to the variable μ. A center unstable manifold $\mathcal{V}^{0+}_{(\mu_0,x_0)}$ corresponding to F is tangent at (μ_0, x_0) to $\mathbf{R} \times V^{0+}_{x_0}$ and is of class C^r. Since $\mathcal{V}^{0+}_{(\mu_0,x_0)}$ is locally attracting for F (see the Center Unstable Manifold Theorem 7.1), all the *locally recurrent* points near (μ_0, x_0) are contained in $\mathcal{V}^{0+}_{(\mu_0,x_0)}$. It suffices, therefore, to study the restriction of the map F to $\mathcal{V}^{0+}_{(\mu_0,x_0)}$.[6] By projecting $\mathcal{V}^{0+}_{(\mu_0,x_0)}$ along $V^{-}_{x_0}$ on the tangent space $\mathbf{R} \times V^{0+}_{x_0}$, we obtain a C^r map $F^* : (\mu, y) \mapsto (\mu, \tilde{f}_\mu(y))$. A fixed point y of $\tilde{f}_\mu$ corresponds to a fixed point X of f_μ, and the eigenvalues of $D_y \tilde{f}_\mu$ are equal to those eigenvalues of $D_x f_\mu$ that are outside of the unit circle or close to it (in $\{z : |z| \geqslant 1\}$ for $D_0 f_\mu$).

We shall be mainly interested in the case when $D_{x_0} f_{\mu_0}$ has its spectrum inside the unit circle apart from one eigenvalue (saddle-node, flip) or two eigenvalues (Hopf). In that case, a center unstable manifold is a center manifold, and $\tilde{f}_\mu$ is a map in one or two dimensions. One may perform the reduction to one or two dimensions more generally by using a center manifold instead of a center unstable manifold.

9.4. General Remarks

(a) The spectrum of a linear operator depends continuously on that operator (see Appendix A.5). In fact, simple eigenvalues of $D_x f_\mu$

---

[6]In fact, F is determined up to *local conjugacy* by its restriction to $\mathcal{V}^{0+}_{(\mu_0,x_0)}$, see Pugh and Shub [1], Palis and Takens [1]. Using this fact, one can study the saddle node, flip, and Hopf bifurcations from the point of view of structural stability (or stability of parametrized families of maps; see Newhouse, Palis, and Takens [1]).

vary smoothly on $\mathcal{F}$ as long as they do not collide. It will be part of our genericity assumptions that, when (μ, x) passes a bifurcation while moving along $\mathcal{F}$, one eigenvalue of $D_x f_\mu$ *crosses* the unit circle at ± 1, or two eigenvalues *cross* at $\alpha, \overline{\alpha}$ (with speed $\neq 0$). Crossing implies that of the two branches of $\mathcal{F}$ (on both sides of the bifurcation point), at most one consists of attracting fixed points. This is shown in Fig. 16 for saddle-node bifurcations.

(b) For each of the three bifurcations considered, there were two cases: *direct* (or *supercritical*) and *inverted* (or *subcritical*) bifurcation (case A and B, respectively, in Figs. 16, 17, and 18). Of course, replacing μ by $-\mu$ replaces inverted bifurcations by direct bifurcations (and the qualification *direct* is generally omitted). Whether a direct or inverted bifurcation occurs depends on higher-order derivatives of f_{μ_0} at x_0 (the sign of a certain expression of second order for the saddle-node bifurcation, of third order for the flip and Hopf bifurcations).

By assumption, x_0 is not hyperbolic and therefore cannot be an attracting fixed point of f_{μ_0} in our terminology. We say that x_0 is *vaguely attracting* if third-order terms make it attracting (saddle nodes are thus excluded); in this case, the bifurcation is supercritical.

(c) Our genericity assumptions will imply that for each of the bifurcations considered, a *square-root law* holds, saying that the diameter of the bifurcated set is $\approx$ const. $\sqrt{\mu - \mu_0}$ (direct bifurcations) or $\approx$ const. $\sqrt{\mu_0 - \mu}$ (inverted bifurcations). The diameter of the bifurcated set is the distance of the two fixed points created at a saddle-node bifurcation, or the distance of the two periodic points created at a flip bifurcation, or the diameter of the invariant circle γ_μ created at a Hopf bifurcation.

(d) Our analysis of bifurcations will characterize the *local recurrent points*. Recurrence may here be taken in the sense of nonwandering or chain recurrence, and local means that we take into account only the orbits that stay in some neighborhood of the bifurcation point.

(e) There is no loss of generality in assuming $\mu_0 = 0$, $x_0 = 0$. For simplicity, we shall state our theorems in that case.

(f) In the statement of theorems, we shall restrict ourselves to the situation where an attracting fixed point is present for $\mu < \mu_0$ (or

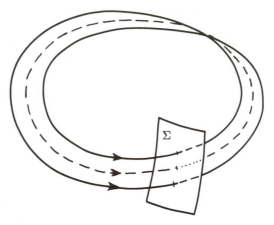

FIG. 20. Period doubling bifurcation for a semiflow, corresponding to a flip bifurcation for its Poincaré map.

$\mu > \mu_0$ for the inverted saddle node). This restriction is not essential but permits more detailed assertions in the cases of greatest practical interest.

10. Bifurcation of Periodic Orbits. The Case of Semiflows

The study of bifurcations of a periodic orbit presents no new problems. The bifurcations of a fixed point for a semiflow require some more care.

10.1. Periodic Orbits

A periodic point of period n for a map f is a fixed point for f^n. The local study of bifurcations of a periodic orbit for a map is thus reduced entirely to the study of bifurcations of a fixed point.

The bifurcations of a periodic orbit for a semiflow also correspond to bifurcations of a fixed point for a map, namely the Poincaré map P. In particular, starting from an orbit of period T, the flip bifurcation will produce a periodic orbit of period $\approx 2T$ (Fig. 20). The Hopf bifurcation will produce a thin 2-torus (Fig. 21).

10.2. Fixed Points for a Semiflow

Let U again be an open subset of the Banach space E, let J be an interval of $\mathbf{R}$, and let $f_\mu^t : U \mapsto E$ be a map defined for each

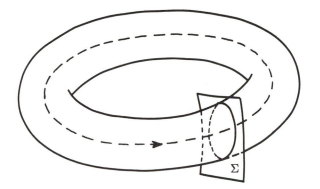

FIG. 21. Creation of an invariant 2-torus for a semiflow, corresponding to a Hopf bifurcation for its Poincaré map.

$\mu \in J$, $t \in [0, T]$, for some $T > 0$. We assume that f_μ^0 is the identity, and $f_\mu^{t_1} \circ f_\mu^{t_2}$ is—where defined—equal to $f_\mu^{t_1 + t_2}$ if $t_1 + t_2 \leqslant T$. It is easy to extend the definition of f_μ^t to $t > T$ (and also $t < 0$ in the case of a diffeomorphism). This is what we shall mean by a *local semiflow*. We assume also that $(\mu, x, t) \mapsto f_\mu^t x$ is continuous on $J \times U \times [0, T]$ and $C^{\mathbf{r}}$ on $J \times U \times (0, T)$, with $\mathbf{r} \geqslant 1$.

Let $(f_{\mu_0}^t)$ have a hyperbolic fixed point x_0, i.e., x_0 is a hyperbolic fixed point for $f_{\mu_0}^\tau$, whenever $\tau \in (0, T)$. For fixed τ, the map f_μ^τ thus has a unique fixed point $x(\mu)$ close to x_0, and $x(\mu)$ is hyperbolic when μ is in a suitable interval $\tilde{J} \ni \mu_0$. If $t \in (0, T - \tau)$, we have $f_\mu^\tau f_\mu^t x(\mu) = f_\mu^t f_\mu^\tau x(\mu) = f_\mu^t x(\mu)$, and therefore $f_\mu^t x(\mu)$ is again a fixed point of f_μ^τ. Since $f_{\mu_0}^t(x_0) = x_0$, the local uniqueness of hyperbolic fixed points implies that $f_\mu^t(x(\mu)) = x(\mu)$ for all $\mu \in \tilde{J}$. Therefore, $x(\mu)$ is a hyperbolic fixed point of (f_μ^t) for all $\mu \in \tilde{J}$.

The same considerations of genericity hold for the fixed points of a (semi)flow as for those of a map. In particular, the set

$$\mathcal{F} = \{(\mu, x) \in J \times U : x \text{ is a fixed point of } (f_\mu^t)\}$$

is generically a nonsingular curve.[7] Bifurcations will occur at $(\mu_0, x_0) \in \mathcal{F}$ when x_0 is not a hyperbolic fixed point for f_{μ_0}, i.e., when $D_{x_0} f_{\mu_0}^1$ is not hyperbolic. Generically, this will occur when $D_{x_0} f_{\mu_0}^1$ has a simple eigenvalue equal to 1 (saddle node), or a pair of complex conjugate

[7]See Problem 1.

eigenvalues of modulus 1 (Hopf bifurcation). A simple eigenvalue at -1 is not possible (not compatible with the fact that $(D_{x_0} f^t_{\mu_0})_{t \geqslant 0}$ is a semigroup of linear transformations); therefore, there is no bifurcation analogous to the flip.

If the local flow (f^t_μ) is obtained by integration of a C^r vector field X_μ (defined in a neighborhood of x_0), the fixed points are defined by $X_\mu(x) = 0$, and x is a hyperbolic fixed point if the spectrum of $D_x X_\mu$ is disjoint from the imaginary axis. At a saddle node, $D_x X_\mu$ has a simple eigenvalue equal to 0. A Hopf bifurcation corresponds to the crossing of the real axis by the two complex conjugate eigenvalues.

11. The Saddle-Node Bifurcation

We discuss successively the case of a map and that of a semiflow.

11.1. Theorem. (Saddle-node for a map). *Let U be a open subset of the Banach space E, J an interval of $\mathbf{R}$, and $(\mu, x) \mapsto f_\mu(x)$ a C^r map $J \times U \mapsto E$, with $\mathbf{r} \geqslant 2$. We assume that $(0,0) \in J \times U$, that $f_0(0) = 0$, and that the spectrum of $D_0 f_0$ is contained in $\{z : |z| < 1\}$, except for a simple eigenvalue at 1. Furthermore we suppose that $\frac{d}{d\mu} f_\mu(0)|_{\mu=0}$ is not contained in the subspace of E corresponding to the part of the spectrum of $D_0 f_0$ in $\{z : |z| < 1\}$ (this is a generic hypothesis).*

Under these conditions the set $\{(\mu, x) : f_\mu(x) = x\}$ is, near $(0,0) \in J \times U$, a 1-dimensional C^r manifold tangent to $(0, u)$ at $(0,0)$, where u is an eigenvector corresponding to the eigenvalue 1 of $D_0 f_0$.[8] In what follows, we restrict $J \times U$ to a suitable neighborhood of $(0,0)$. There are two generic possibilities, depending on the sign of some coefficient computed from second-order derivatives of f_0.

(A) (Direct bifurcation). *The map f_μ has no fixed point for $\mu < 0$, one fixed point at 0 for $\mu = 0$, and two hyperbolic fixed points for $\mu > 0$ (of which one is attracting).*

(B) (Inverted bifurcation). *The map f_μ has two hyperbolic fixed points for $\mu < 0$ (of which one is attracting), one fixed point at 0 for $\mu = 0$, and no fixed point for $\mu > 0$.*

In both cases, the distance of the fixed points to 0 is $O(|\mu|^{1/2})$ (i.e., of

---

[8]For this result, it would suffice to assume $\mathbf{r} \geqslant 1$.

order $\sqrt{|\mu|}$), and there are no local recurrent points other than the fixed points indicated above.

By the implicit function theorem, we know that $\{(\mu, x) : f_\mu(x) = x\}$ is a $C^{\mathbf{r}}$ manifold tangent to $\{0\} \times V_0^0$, where V_0^0 is the eigenspace of $D_0 f_0$ corresponding to the eigenvalue 1. Using a C^2 center manifold as explained in Section 9.3, we replace f_μ by $\tilde{f}_\mu : \tilde{U} \mapsto \mathbf{R}$, where $\tilde{U}$ is a neighborhood of 0 in $\mathbf{R}$. We may write

(11.1) $$\tilde{f}_\mu(y) = a\mu + y + by^2 + o(|\mu| + y^2),$$

where $a \neq 0$ by assumption, and $o(|\mu| + y^2)$ is a term depending on μ, y such that $o(|\mu| + y^2)/(|\mu| + y^2) \to 0$ when $\mu, y \to 0$. The manifold $\{(\mu, y) : \tilde{f}_\mu(y) = y\}$ is thus the graph of a map $y \to \mu^*(y)$ such that

(11.2) $$\mu^*(y) = -\frac{b}{a}y^2 + o(y^2).$$

By making the generic hypothesis that $b \neq 0$, we shall obtain case (A) or (B) of the theorem depending on whether $b/a < 0$ or $b/a > 0$. Notice that, by continuity, the derivative of the map $(\mu, y) \mapsto \tilde{f}_\mu(y) - y$ does not vanish near $(0, 0)$. This derivative is a linear function $(\mu', y') \mapsto A(\mu, y)\mu' + B(\mu, y)y'$. Because $b/a \neq 0$ in (11.2), we see that $B(\mu, y)$ does not vanish when $\tilde{f}_\mu(y) = y$, except at $(0, 0)$. This implies that the fixed points of $\tilde{f}_\mu$ are hyperbolic for $\mu \neq 0$. To relate the behavior of f_μ to that of $\tilde{f}_\mu$, the locally attracting character of the center unstable manifold is used. The details are easy and are left to the reader.

11.2. Theorem (Saddle-node for a semiflow). *Let U be an open subset of a Banach space E, J an interval of $\mathbf{R}$, and $(\mu, x, t) \mapsto f_\mu^t x$ a continuous map $J \times U \times [0, T] \mapsto E$ that is $C^{\mathbf{r}}$ on $J \times U \times (0, T)$, with $\mathbf{r} \geqslant 2$. We assume that this map defines a local semiflow, that $(0, 0) \in J \times U$, that $f_0^t(0) = 0$, and that for $t \in (0, T)$, the spectrum of $D_0 f_0^t$ is contained in $\{z : |z| < 1\}$ except for a simple eigenvalue at 1. Furthermore, for some $\tau \in (0, T)$, we suppose that $\frac{d}{d\mu} f_\mu^\tau(0)|_{\mu=0}$ is not contained in the subspace of E corresponding to the part of the spectrum of $D_0 f_0^\tau$ in $\{z : |z| < 1\}$ (this is a generic hypothesis).*

Under these conditions, the set $\{(\mu, x) : F_\mu^t x = x \text{ for all } t \in [0, T]\}$ is a one-dimensional $C^{\mathbf{r}}$ manifold near $(0, 0) \in J \times U$,[9] and the other

---

[9] For this result, it would suffice to assume $\mathbf{r} \geqslant 1$.

conclusions of Theorem 11.1 apply, with the simple replacement of f_μ by (f_μ^t).

By assumption, $f_\mu = f_\mu^\tau$ satisfies the assumptions of Theorem 11.1 for some $\tau \in (0, T)$, and therefore its conclusions. If x is a fixed point for f_μ, it is a fixed point for (f_μ^t) because a τ-periodic orbit near 0, contained in the 1-dimensional manifold $\{(\mu, x) : f_\mu x = x\}$, must reduce to a point. In the generic cases (A), (B), the fixed points obtained for $\mu \neq 0$ are hyperbolic for the semiflow, and one of them is attracting (see Section 5.3). Apart from the fixed points, there are no local recurrent points: If $f_\mu^t x$ remains near 0, it tends to a fixed point by Theorem 11.1.

11.3. Bursts of Turbulence. The Manneville–Pomeau Phenomenon

If a dynamical system undergoes an inverted saddle-node bifurcation, the system will be observed in a state corresponding to an attracting fixed point when $\mu < \mu_0$. The behavior for $\mu > \mu_0$ cannot be determined from local considerations alone. It may be that the point describing the system jumps from x_0 to a distant attractor. It may also happen that a new attractor A_μ is created, such that $\lim A_\mu \ni x_0$ when $\mu \to \mu_0$. For μ close to μ_0, an orbit $\{f^k x\}$ on the attractor will then spend much time near x_0, which is still almost a fixed point (we may call x_0 a *phantom fixed point*). This situation is of particular interest for the Poincaré map of a semiflow. In that case, the periodic motion visible for $\mu < \mu_0$ still seems present for $\mu > \mu_0$, but interrupted by excursions on the attractor, away from the *phantom periodic orbit*. These excursions are called *bursts of turbulence* in the case where A_μ is a *strange attractor* (see Fig. 22), but the same phenomenon occurs with nonstrange attractors.[10] For μ close to μ_0, the bursts are infrequent, but they become more and more prevalent as μ increases. Such a bifurcation from periodicity to the aperiodic motion on a strange attractor has been termed *intermittent transition to turbulence* by Pomeau and Manneville [1]. It is one way in which *chaos* can arise in dynamical systems associated with natural phenomena.

Figure 23 shows the graph of the 1-dimensional map $\tilde{f}_\mu$ of equation (11.1) with $a > 0, b > 0$. For small $\mu > 0$, how many consecutive points on an orbit can there be in an interval $(-c, +c)$ around 0? Notice that $\tilde{f}_\mu$ is, up to higher-order terms, the time one map of the flow defined by

---

[10]See Section 13.4.

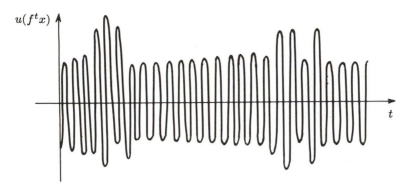

$$F_{IG.}\ 22.$$ Bursts of turbulence in an apparently periodic background. The time evolution f^t is on a strange attractor obtained by bifurcation from an attracting periodic orbit; u is a real-valued function on the attractor.

the vector field

$$y \mapsto \tilde{X}_\mu(y) = a\mu + by^2.$$

Integrating the equation

$$\frac{dy}{a\mu + by^2} = dt,$$

i.e.,

$$d[\arctan(y(b/a\mu)^{1/2})] = (ab\mu)^{1/2}dt,$$

we find that the time needed to go from $-c$ to c is $n \approx \pi(ab\mu)^{-1/2}$. Allowing again $\mu_0 \neq 0$ we see that n is $O((\mu - \mu_0)^{-1/2})$. In the case of a semiflow, the bursts of turbulence will thus be separated by approximately periodic intervals with duration of the order of $(\mu - \mu_0)^{-1/2}$.

12. The Flip Bifurcation

This bifurcation occurs for maps, but not for semiflows. The map f_μ has a fixed point $x(\mu)$ depending smoothly on μ. We shall, by a change of coordinate, take $x(\mu) = 0$ independently of μ.

12.1. Theorem (Flip for a map). *Let U be an open subset of the Banach space E, J an interval of $\mathbf{R}$, and $(\mu, x) \mapsto f_\mu(x)$ a $C^{\mathbf{r}}$ map $J \times U \mapsto E$, with $\mathbf{r} \geqslant 3$. We assume that $(0,0) \in J \times U$, that $f_\mu(0) = 0$, and that the spectrum of $D_0 f_\mu$ is contained in $\{z : |z| < 1\}$ except for a simple real eigenvalue α_μ such that*

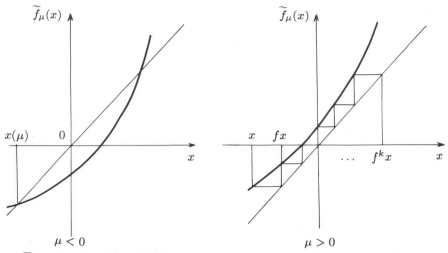

FIG. 23.   The Manneville–Pomeau phenomenon: an attracting fixed point disappears in an inverted saddle-node bifurcation, but orbits remain a long time near 0.

(a) $\alpha_0 = -1$,

(b) $\frac{d}{d\mu}\alpha_\mu < 0$.

Under these conditions the set $\{(\mu, x) : f_\mu^2(x) = x\}$ consists, near $(0,0) \in J \times U$, of $J \times \{0\}$ and a 1-dimensional C^{r-1} manifold tangent to $(0, u)$ at $(0,0)$, where u is an eigenvector corresponding to the eigenvalue -1 of $D_0 f_0$.[11] *In what follows, we restrict $J \times U$ to a suitable neighborhood of $(0,0)$. There are two generic possibilities depending on the sign of some coefficient computed from derivatives up to order 3 of f_0. In both cases, 0 is an attracting fixed point for f_μ when $\mu < 0$.*

(A) (Direct bifurcation: 0 is a vague attractor for f_0).   *The map f_μ has an attracting periodic orbit of period 2 for $\mu > 0$.*

(B) (Inverted bifurcation).   *The map f_μ has a (nonattracting) hyperbolic periodic orbit of period 2 for $\mu < 0$.*

In both cases, the distance of the periodic points to 0 is $O(|\mu|^{1/2})$, and there are no local recurrent points other than the fixed point 0 and the periodic points indicated above.

Let thus u be an eigenvector of $D_0 f_0$ corresponding to the eigenvalue

---

[11] For this result, it would suffice to assume $r \geqslant 2$.

-1, and F be the subspace of E corresponding to the part of the spectrum of $D_0 f_0$ in $\{z : |z| < 1\}$. A $C^{\mathbf{r}-1}$ map Φ of a neighborhood of $(0,0,0)$ in $\mathbf{R} \times \mathbf{R} \times F$ to E is defined by

$$\Phi(\mu, y, x') = \int_0^1 d\tau [D_{\tau y(u+x')} f_\mu^2](u + x') - (u + x').$$

When $y \neq 0$, we thus have

$$\Phi(\mu, y, x') = y^{-1} f_\mu^2(y(u + x')) - (u + x').$$

Notice also that $\Phi(0,0,0) = 0$. The derivative of $\Phi(\mu, y, x')$ with respect to μ is

$$\frac{d}{d\mu}(D_0 f_\mu^2) u|_{\mu=0} = \left(\frac{d\alpha_\mu^2}{d\mu}\right)_{\mu=0} \cdot u - (D_0 f_0^2 - 1) \cdot \left(\frac{du_\mu}{d\mu}\right)_{\mu=0},$$

where u_μ is an eigenvector of $D_0 f_\mu$ depending smoothly on μ, and such that $u_0 = u$. The derivative of Φ with respect to x' at $(0,0,0)$ is $(D_0 f_0^2 - 1)$. Since the derivatives of Φ with respect to μ and x' at $(0,0,0)$ span E, the implicit function theorem yields $C^{\mathbf{r}-1}$ functions $y \mapsto \mu, x'$ such that $\Phi(\mu, y, x') = 0$ near $(0,0,0)$, or equivalently, $f_\mu^2(y(u+x')) = y(u+x')$. This shows that $\{(\mu, x) : f_\mu^2(x) = x\}$ consists of $J \times \{0\}$ and a 1-dimensional $C^{\mathbf{r}-1}$ manifold.

By using a C^3 center manifold (see Section 9.3) we replace f_μ by $\tilde{f}_\mu$: $\tilde{U} \mapsto \mathbf{R}$ with $0 \in \tilde{U} \subset \mathbf{R}$. The center manifold $V^0_{(0,0)} \subset J \times U$ contains the fixed points $(\mu, 0)$ because they are local recurrent points. By local invariance of the center manifold, the eigenvector u_μ corresponding to α_μ is tangent to $V^0_{(0,0)}$ at $(\mu, 0)$, and therefore

$$D_0 \tilde{f}_\mu = \alpha_\mu,$$
$$\tilde{f}_\mu(y) = -y + a\mu y + by^2 + cy^3 + o(|\mu y| + |y|^3),$$

where $a = \frac{d\alpha_\mu}{d\mu}|_{\mu=0} < 0$. We perform the change of variable $\mathbf{y} = y - \frac{1}{2}by^2$, replacing $\tilde{f}$ by the *normal form* $\mathbf{f}$:

$$\mathbf{f}_\mu(\mathbf{y}) = -\mathbf{y} + a\mu\mathbf{y} + c\mathbf{y}^3 + o(|\mu\mathbf{y}| + |\mathbf{y}|^3)$$

so that

$$\mathbf{f}_\mu^2(\mathbf{y}) = \mathbf{y} - 2a\mu\mathbf{y} - 2c\mathbf{y}^3 + o(|\mu\mathbf{y}| + |\mathbf{y}|^3).$$

Generically, $c \neq 0$, and $c > 0$ if 0 is a vague attractor for f_0 (because $f_0(\mathbf{y}) = \mathbf{y}(-1 + c\mathbf{y}^2) + \cdots$). The function $(\mu, \mathbf{y}) \mapsto \mathbf{y}^{-1}[\mathbf{f}^2(\mathbf{y}) - \mathbf{y}]$ is $C^{\mathbf{r}-1}$ and equal to $-2a\mu - 2c\mathbf{y}^2 + o(|\mu| + \mathbf{y}^2)$. By using this, and by proceeding as for Theorem 11.1, one can readily complete the proof of the theorem.

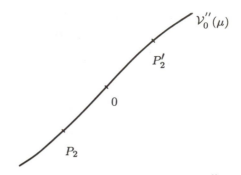

FIG. 24.   Flip bifurcation: the center unstable manifold $V_0''(\mu)$ of 0 contains the periodic points P_2, P_2', and the segment $P_2 P_2'$ is the unstable manifold of 0.

12.2. Remark on the Invariant Manifolds

(a) The center unstable manifold $V_{(0,0)}^0 = V_{(0,0)}^{0+}$ considered in the proof intersects the space $\mu =$ constant in a 1-dimensional manifold $V_0''(\mu)$, which is a center unstable manifold for f_μ (see Theorem 7.1; we assume $|\mu|$ sufficiently small). Suppose for definiteness that we have a direct flip bifurcation. Then $V_0''(\mu)$ contains the points of period 2 for $\mu > 0$ (see Fig. 24). The piece of $V_0''(\mu)$ contained between the periodic points is the (global) unstable manifold of 0. For this piece, we have thus uniqueness, but uniqueness does not extend beyond the periodic points.

Let $P_\mu : \Sigma \mapsto \Sigma$ be a Poincaré map for a semiflow (f_μ^t). Suppose that (f_μ^t) has an attracting periodic orbit of period $\approx T$ for $\mu < \mu_0$, replaced by an attracting periodic orbit of period $\approx 2T$ for $\mu > \mu_0$ as a consequence of a flip bifurcation of P_μ (see Fig. 20). The new periodic orbit is the boundary of the unstable manifold of the periodic orbit of period $\approx T$, which persists for $\mu > \mu_0$ (but is no longer attracting). Notice that this unstable manifold is a *Moebius band*.

12.3. The Feigenbaum Bifurcation

Flip bifurcations for a map may occur in succession and accumulate to a limit, with an attracting fixed point successively being replaced by periodic orbits of period 2^n for $n = 1, 2, \cdots \to \infty$. The resulting *period-doubling cascade* is a new bifurcation known as *Feigenbaum bifurcation*, and there is a corresponding bifurcation for periodic orbits of a semiflow. A detailed study of the Feigenbaum bifurcation is beyond the scope of this monograph; we shall here limit ourselves to some general considerations and references to the literature.

Feigenbaum's original study concerned one-parameter families (f_μ) of one-humped maps of the interval $[0, 1]$.[12] A typical family is given by

$$(12.1) \qquad f_\mu x = 1 - \mu x^2$$

for $0 < \mu \leqslant 2$. A numerical study of this family shows, as μ increases, a sequence of bifurcation points μ_n, where an attracting periodic orbit of period 2^{n-1} is replaced by an attracting periodic orbit of period 2^n. By writing $\mu_\infty = \lim \mu_n$ one finds that $\mu_\infty = 1.401155\ldots$ for the family (12.1), and that

$$(12.2) \qquad \lim_{n \to \infty} \frac{\mu_\infty - \mu_{n-1}}{\mu_\infty - \mu_n} = \delta = 4.66920\ldots.$$

Of course, μ_∞ depends on the choice of the family (f_μ) but—remarkably—δ is *universal* in the sense that many different families give the same value of δ.

To understand the period-doubling cascade, Feigenbaum reflected that the iterate $f_\mu^{2^n}$ must behave like $f_\mu^{2^{n-1}}$, up to rescaling. This is indeed a natural point of view for a physicist familiar with *renormalization group* ideas, and it led him to study the map

$$(12.3) \qquad \left.\begin{array}{c} \Psi : \psi(x) \mapsto -\frac{1}{a}\psi \cdot \psi(-ax) \\ \text{with } a = -\psi(1) \end{array}\right\},$$

acting on a space of functions $\psi : [-1, 1] \mapsto [-1, 1]$ such that $\psi(0) = 1$.[13] Numerically, Feigenbaum discovered a fixed point g of Ψ. The function g is even and satisfies the Cvitanović–Feigenbaum equation

$$(12.4) \qquad g \circ g(ax) + ag(x) = 0.$$

It turns out that g is a hyperbolic fixed point of (12.3) and that the spectrum of the derivative of Ψ at g is contained in $\{z : |z| < 1\}$, apart

---

[12] See Feigenbaum [1], [2]. A lot is known about maps of the interval; for a general presentation of the subject, see the monograph of Collet and Eckmann [1].

[13] We do not further specify this space here. In fact, Feigenbaum proceeded a little differently, and considered a map $\tilde{\Psi} : \psi(x) \mapsto -\alpha\psi \circ \psi(-x/\alpha)$, where α is determined by the nonlinear eigenvalue equation (12.4): $\alpha = a^{-1} = (-g(1))^{-1} = 2.5029078750957\ldots.$

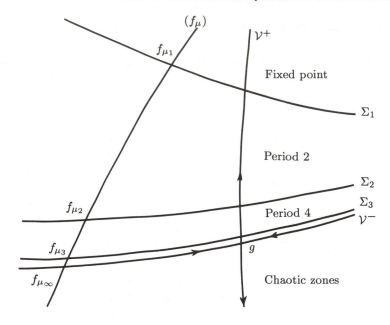

FIG. 25. The fixed point of the Feigenbaum map Φ, and its stable and unstable manifolds $\mathcal{V}^-$, $\mathcal{V}^+$. Creation of an attracting orbit of period 2^n from an attracting orbit of 2^{n-1} occurs at the bifurcation manifold Σ_n; Φ maps Σ_n to Σ_{n-1} and the Σ_n pile up on $\mathcal{V}^-$. The successive bifurcation points for a one-parameter family (f_u) are also shown.

from a simple eigenvalue $\delta = 4.66920\ldots$ (we shall see in a moment, why this is the same number that appears in (12.2)). The unstable and stable manifolds $\mathcal{V}^+$, $\mathcal{V}^-$ of g thus have dimension 1 and codimension 1, respectively.

We shall now consider a subset Σ_1 of the bifurcation set (see Section 8.6), where an attracting fixed point is replaced by an attracting period 2 orbit (flip bifurcation). One can choose Σ_1 so that it is a smooth codimension-1 manifold intersecting $\mathcal{V}^+$ transversally (see Fig. 25). Consider now the manifolds $\Sigma_k = (\Phi^{k-1})^{-1}\Sigma_1$ (restricted to a suitable neighborhood of g). It is easy to check that at the crossing of Σ_k, an attracting orbit of period 2^{k-1} is replaced by an attracting orbit of period 2^k. When $k \to \infty$, the manifolds Σ_k pile up on $\mathcal{V}^-$ in a way described by the *inclination lemma* (Problem 5 of Part 1). Roughly speaking, the Σ_n become parallel to $\mathcal{V}^-$, with the distance to $\mathcal{V}^-$ decreasing geometrically like δ^{-n}.

Let now (f_μ) be a smooth one-parameter family of maps that crosses $\mathcal{V}^-$ transversally (with nonzero speed) for $\mu = \mu_\infty$. One can then define μ_k, for sufficiently large k, such that $f_{\mu_k} \in \Sigma_k$, and $\mu_k \to \mu_\infty$ when $k \to \infty$. The consequences of the inclination lemma referred to above then yield (12.2). This shows that the manifold $\mathcal{V}^-$ is the bifurcation hypersurface for a codimension-1 bifurcation (the Feigenbaum bifurcation), and explains the *universality* of the limit (12.2). Arbitrarily close to $\mathcal{V}^-$, on the side opposite to the manifolds Σ_n, one can show that there are zones with sensitive dependence on initial condition. The Feigenbaum bifurcation, therefore, is one of the ways in which a dynamical system may become *chaotic*.

One can extend this analysis to dimensions greater than 1 and obtain a theory of the Feigenbaum bifurcation for maps (this was done rigorously by Collet, Eckmann, and Koch [1]). The extension to semiflows is simply obtained by suspension.

But let us go back to the problem of obtaining proofs in the one-dimensional situation. How does one establish the existence of the fixed point g, estimate the spectrum of $D_g\Psi$, and prove that $\mathcal{V}^+$ intersects Σ_1 transversally? A rigorous treatment of these questions is relatively hard, although the principle is simple. One uses the stability of hyperbolic fixed points and transversal intersections under perturbations. In particular, if one has a good approximation of Ψ, with good estimates of derivatives, and a good approximate fixed point, one knows that there is a unique true fixed point g near the approximate one, and one has estimates of the spectrum of $D_g\Psi$. Lanford [2] has implemented this program by letting Ψ act on a space of holomorphic functions and approximating g by a polynomial. In this approach, then, the problem boils down to verifying complicated numerical estimates. In fact, the computations involved are sufficiently hard that it is impractical to do them by hand. Lanford's proof is thus a *computer-aided proof*, where the computer is used to check some complicated numerical inequalities.[14] Let us stress that this approach is perfectly rigorous. Its main imperfection is that it is at present limited to analytic functions: One would like a theorem establishing the Feigenbaum bifurcation for $C^\mathbf{r}$ functions with some $\mathbf{r} < \infty$.

---

[14]To prove only the existence of a solution g of the nonlinear eigenvalue problem (12.4) is easier, see Campanino, Epstein, and Ruelle [1], Campanino and Epstein [1], Epstein [1].

13. The Hopf Bifurcation

This bifurcation occurs for maps and for flows, and there are some important differences between the two cases. The map f_μ or the flow (f_μ^t) has a fixed point $x(\mu)$ depending smoothly on μ. We shall, by a change of coordinate, take $x(\mu) = 0$ independently of μ. The concept of a normally hyperbolic invariant manifold used in the following theorem is explained below and discussed in Section 14.

13.1. Theorem (Hopf for a map).[15] *Let U be an open subset of the Banach space E, J an interval of $\mathbf{R}$, and $(\mu, x) \mapsto f_\mu(x)$ a C^2 map $J \times U \mapsto E$, such that f_μ is $C^{\mathbf{r}}$ for each μ, with $3 \leqslant \mathbf{r} < \infty$. We assume that $(0,0) \in J \times U$, that $f_\mu(0) = 0$, and that the spectrum of $D_0 f_\mu$ is contained in $\{z : |z| < 1\}$, except for two eigenvalues α_μ, $\bar{\alpha}_\mu$ such that*

(a) $|\alpha_0| = 1$, $\alpha_0^3 \neq 1$, $\alpha_0^4 \neq 1$

(b) $\frac{d}{d\mu}|\alpha_\mu| > 0$.

We also make the generic assumption that some coefficient C computed from derivatives up to order 3 of f_0 does not vanish, and we restrict $J \times U$ to a suitable neighborhood of $(0,0)$.

Under these conditions, 0 is an attracting fixed point for f_μ when $\mu < 0$. For $\mu > 0$ or $\mu < 0$, depending on the sign of C, f_μ also has an invariant manifold H_μ (i.e., $f_\mu H_\mu = H_\mu$). The manifold H_μ is $C^{\mathbf{r}}$ diffeomorphic to a circle, and consists of points at distance $O(|\mu|^{1/2})$ of 0. If f_μ depends continuously on μ with respect to the $C^{\mathbf{r}}$ topology, so does H_μ.

(A)(Direct bifurcation: 0 is a vague attractor for f_0). *H_μ is an attracting ($\mathbf{r}$-normally hyperbolic) invariant manifold, present for $\mu > 0$.*

(B) (Inverted bifurcation). *H_μ is a (nonattracting) $\mathbf{r}$-normally hyperbolic invariant manifold, present for $\mu < 0$.*

Each compact invariant manifold close to 0 is contained in $H_\mu \cup \{0\}$ (uniqueness of H_μ). The local recurrent points are the fixed point 0 and some points of H_μ.[16]

---

[15]The case of flows (Theorem 13.3) was studied first: by Poincaré, Andronov (in 2 dimensions) and Hopf [1]. The case of diffeomorphisms was discussed by Neĭmark [1], Sacker [1] (they consider the bifurcation of a periodic orbit for a flow), Ruelle and Takens [1]. For a comprehensive discussion and references see Marsden and McCracken [1].

[16]See Section 13.2.

To prove this theorem, we shall proceed via several successive changes of variables. By using a center manifold (see Section 9.3), we first replace f_μ by $\tilde{f}_\mu : \tilde{U} \mapsto \mathbf{R}^2$, with $0 \in \tilde{U} \subset \mathbf{R}^2$. The center manifold $V^0_{(0,0)} \subset J \times U$ contains the fixed points $(\mu, 0)$ because they are local recurrent points. By construction of the center manifold (Section 7), we may assume that $(\mu, \cdot) \mapsto \tilde{f}_\mu \cdot$ is C^2, and that each $\tilde{f}_\mu$ is $C^{\mathbf{r}}$. ($\tilde{f}_\mu$ depends continuously on μ for the $C^{\mathbf{r}}$ topology if f_μ does.) If $u_\mu, \bar{u}_\mu$ are complex eigenvectors corresponding to $\alpha_\mu, \bar{\alpha}_\mu$, the real vectors $u'_\mu = \frac{1}{2}(u_\mu + \bar{u}_\mu)$, $u''_\mu \frac{1}{2i}(u_\mu - \bar{u}_\mu)$ are tangent to $V^0_{(0,0)}$ at $(\mu, 0)$ by local invariance of the center manifold. Therefore,

$$D_0 \tilde{f}_\mu = A_\mu^{-1} \begin{pmatrix} \operatorname{Re} \alpha_\mu & -\operatorname{Im} \alpha_\mu \\ \operatorname{Im} \alpha_\mu & \operatorname{Re} \alpha_\mu \end{pmatrix} A_\mu,$$

where A_μ is a 2×2 matrix with C^1 dependence on μ, and $A_0 = 1$. Let $\hat{f}_\mu = A_\mu \circ \tilde{f}_\mu \circ A_\mu^{-1}$. By identifying $\mathbf{R}^2$ with $\mathbf{C}$, we see that

$$\hat{f}_\mu(z) = \alpha_0 z + a\mu z + \mathcal{B}(z, \bar{z}) + \mathcal{C}(z, \bar{z}) + \mathcal{R}(\mu, z, \bar{z}).$$

Here $a = \frac{d\alpha_\mu}{d\mu}\big|_{\mu=0}$, and $\mathcal{B}, \mathcal{C}$ are a quadratic and a cubic form in z, $\bar{z}$. The remainder term $\mathcal{R}$ satisfies $\mathcal{R} = o(|\mu z| + |z^3|)$, i.e., $\mathcal{R}/(|\mu z| + |z^3|) \to 0$ when $\mu, z \to 0$, and corresponding estimates for the derivatives.

Our dynamical system will be more manageable if we put $\hat{f}$ in a *normal form* $\mathbf{f}$ by a polynomial change of coordinates $z \mapsto \mathbf{z} = \psi(z) = z + \psi_2 + \psi_3$ where ψ_m is homogeneous of degree m in $z, \bar{z}$. Term by term, identification in the relation

$$\psi((\alpha_0 + a\mu)z + \mathcal{B}(z, \bar{z}) + \mathcal{C}(z, \bar{z}) + o(|\mu z| + |z^3|)) = \mathbf{f}(\psi(z))$$

permits the elimination of quadratic terms in $\mathbf{f}$, and leaves only a $\mathbf{z}(\bar{\mathbf{z}}\mathbf{z})$ cubic term. We can indeed solve

$$(\alpha_0 + a\mu)z + \mathcal{B}(z, \bar{z}) + \mathcal{C}(z, \bar{z}) + \psi_2(\alpha_0 z) + \psi_3(\alpha_0 z)$$
$$= (\alpha_0 + a\mu)(z + \psi_2(z) + \psi_3(z)) + cz(\bar{z}z)$$
$$+ \text{ higher order}$$

because, when $p + q = 2$ or 3, $\alpha_0^p \bar{\alpha}_0^q - \alpha_0 \neq 0$ except for $p = 2$, $q = 1$ (Assumption (a)). This determines completely $\mathbf{f}$ and ψ_2, ψ_3, except for the $z(\bar{z}z)$ term of ψ_3, which may be chosen to vanish. The normal form is thus

(13.1)
$$\begin{aligned}
\mathbf{f}_\mu(\mathbf{z}) &= \alpha_0 \mathbf{z} + a\mu\mathbf{z} + c\mathbf{z}(\bar{\mathbf{z}}\mathbf{z}) + R(\mu, \mathbf{z}) \\
&= \alpha_0 \mathbf{z} \exp[(\bar{\alpha}_0 a)\mu + (\bar{\alpha}_0 c)(\bar{\mathbf{z}}\mathbf{z})] + R_1(\mu, \mathbf{z}).
\end{aligned}$$

Therefore,

(13.2) $|\mathbf{f}_\mu(\mathbf{z})|^2 = |\mathbf{z}|^2 \exp[2A\mu + 2C\bar{\mathbf{z}}\mathbf{z}] + R_2(\mu, \mathbf{z}),$

where $A = \mathrm{Re}(\bar{\alpha}_0 a)$, $C = \mathrm{Re}(\bar{\alpha}_0 c)$, and $A > 0$ by Assumption (b). The remainder terms in (13.1), (13.2) satisfy

$$\begin{aligned}
R &= \mathrm{o}(|\mu\mathbf{z}| + |\mathbf{z}^3|), \qquad R_1 = \mathrm{o}(|\mu\mathbf{z}| + |\mathbf{z}^3|), \\
R_2 &= \mathrm{o}(|\mu\mathbf{z}^2| + |\mathbf{z}^4|),
\end{aligned}$$

and correspondingly for derivatives.

We make the generic assumption that $C \neq 0$. For definiteness, we shall discuss the case $C < 0$, and we write

$$\mathbf{z} = \left(\frac{A\mu}{-C}\right)^{1/2} e^{\eta + i\theta}$$

for $\mu > 0$. We now have a dynamical system

$$(\eta, \theta) \mapsto (g_\mu(\eta, \theta), h_\mu(\eta, \theta)),$$

and an easy calculation yields

(13.3)
$$g_\mu(\eta, \theta) = \eta\left(1 - 2A\mu\frac{e^{2\eta} - 1}{2\eta}\right) + \mathrm{o}_1,$$

(13.4)
$$h_\mu(\eta, \theta) = \theta_0 + \theta + E\mu + F\mu e^{2\eta} + \mathrm{o}_2 \qquad (\mathrm{mod}\, 2\pi).$$

We have written $\alpha_0 = \exp\theta_0$. We shall assume $|\eta| < K$, and since η is bounded, we may write

$$\mathrm{o}_1 = \mathrm{o}(\mu), \qquad \mathrm{o}_2 = \mathrm{o}(\mu),$$

and correspondingly for the derivatives.

Let $\psi : \theta \mapsto \eta$ be a C^1 map and (ψ_n) the sequence of its successive graph transforms. We thus have

$$(13.5) \qquad \psi_n(h(\psi_{n-1}(\theta), \theta)) = g(\psi_{n-1}(\theta), \theta).$$

In view of (13.3), given K, there is μ_0 such that if $|\psi| \leqslant K$ and $\mu \leqslant \mu_0$, then $|\psi_n| \leqslant K$ for all n. In fact, for suitable $N(\mu)$, a bound $|\psi_n| \leqslant \varepsilon(\mu)$ such that $\lim_{\mu \to \infty} \varepsilon(\mu) = 0$ holds for $n \geqslant N(\mu)$. By differentiating (13.5), we obtain

$$(13.6) \qquad \begin{aligned} \psi_n'(\theta_n) \cdot [(2F\mu e^{2\psi_{n-1}} + \mathrm{o}_3)\psi_{n-1}' + 1 + \mathrm{o}_4 \\ = \psi_{n-1}'(\theta_{n-1})(1 - 2A\mu e^{2\psi_{n-1}} + \mathrm{o}_5) + \mathrm{o}_6, \end{aligned}$$

where o_3, o_4, o_5, o_6 are $\mathrm{o}(\mu)$ and $\theta_n = h(\psi_{n-1}(\theta_{n-1}), \theta_{n-1})$. Let now $K_1 \leqslant A/3FK$; there is $\mu_1 \in (0, \mu_0]$ such that $|\psi'| \leqslant K_1$ for $\mu \leqslant \mu_1$ implies $|\psi_n'| \leqslant K_1$ for all n. Furthermore, for suitable $N_1(\mu) \geqslant N_0(\mu)$, a bound $|\psi_n'| \leqslant \varepsilon_1(\mu)$ such that $\lim_{\mu \to 0} \varepsilon_1(\mu) = 0$ holds for $n \geqslant N_1(\mu)$.

Notice that (13.3), (13.4) yield

$$(13.7)$$
$$g_\mu(\eta, \theta) - g_\mu(\eta^*, \theta) = (\eta - \eta^*)\left(1 - 2A\mu \frac{e^{2\eta} - e^{2\eta^*}}{2\eta - 2\eta^*} + \mathrm{o}(\mu)\right),$$

$$(13.8)$$
$$h_\mu(\eta, \theta) - h_\mu(\eta^*, \theta) = (\eta - \eta^*)\left(2F\mu \frac{e^{2\eta} - e^{2\eta^*}}{2\eta - 2\eta^*} + \mathrm{o}(\mu)\right).$$

These estimates, together with $|\psi_n'| \leqslant \varepsilon_1(\mu)$ and $|\psi_n| \leqslant \varepsilon(\mu)$ imply that ψ_n converges uniformly to a limit φ when $\mu \leqslant \mu_2$, μ_2 small enough. (The convergence is exponential, but not uniformly so with respect to μ.) The limit φ is independent of the choice of ψ. Using (13.6) and $K_1 \leqslant A/3FK$, it is now easy to see that the derivatives ψ_n' converge uniformly to a limit, which is necessarily φ'. The functions $(\mu, \theta) \mapsto \varphi, \varphi'$ are thus continuous, and $|\varphi| \leqslant \varepsilon(\mu)$, $|\varphi'| \leqslant \varepsilon_1(\mu)$.

The image of $[0, 2\pi]$ by the map

$$\theta \to \left(\frac{A\mu}{-C}\right)^{1/2} e^{\varphi(\theta) + i\theta}$$

is a C^1 curve H_μ, diffeomorphic to a circle. It is our desired invariant manifold. If we set $\eta^* = \varphi(\theta)$ in (13.7), (13.8) and use our estimate on

φ', we see that H_μ attracts the orbits $(\mathbf{f}_\mu^n \mathbf{z})$ for $|\mathbf{z}|^2 \in (\frac{A\mu}{-C}e^{-2K}, \frac{A\mu}{-C}e^{2K})$. If $|\mathbf{z}|^2$ is not in that interval, or if $\mu \leqslant 0$, the behavior of the orbit is easily obtained from (13.2), provided that $(\mu, \mathbf{z})$ remains in a small neighborhood of $(0,0)$ in $\mathbf{R} \times \mathbf{C}$. The statements about local recurrent points in the theorem are thus readily verified.

Equations (13.7) and (13.8) (or Eq. (13.1)) show that, near H_μ, the map $\mathbf{f}_\mu$ compresses radial increments of $\mathbf{z}$ by a factor $\approx 1-$ const. $|\mathbf{z}|^2$. On the other hand, an angular increment of $\mathbf{z}$, since it does not change $|\mathbf{z}|^2$, will be multiplied by a factor $\approx 1+\mathrm{o}(|\mathbf{z}|^2)$. We have here a situation known as *normal hyperbolicity*: A map f leaves a submanifold invariant, and f contracts or expands in transversal directions more than it distorts lengths on the submanifold. We shall study normal hyperbolicity in Section 14. Here we only note that if a C^1 submanifold H is normally hyperbolic sufficiently strongly under a $C^\mathbf{r}$ map f (more precisely if it is *r-normally hyperbolic*), then the submanifold H is actually of class $C^\mathbf{r}$ and depends continuously on f with respect to the $C^\mathbf{r}$ topologies of H and of f (Theorem 14). In particular, H_μ is of class $C^\mathbf{r}$ and depends continuously on μ with respect to the $C^\mathbf{r}$ topology. We can now go back from the variable $\mathbf{z}$ to the original x, and this completes the proof of the theorem.

Before turning to the Hopf bifurcation for semiflows (Section 13.3), we shall now digress to the important subject of circle maps.

13.2. Diffeomorphisms of the Circle

The Hopf bifurcation for a map produces an invariant circle H_μ for $\mu > 0$, or $\mu < 0$, and f_μ restricted to H_μ is a diffeomorphism. We are thus led to discussing diffeomorphisms of the circle, but this is a vast subject, and its serious study requires delicate techniques (small denominators); we shall thus limit ourselves here to a general description of some important facts.

The circle S^1 may be identified with $\mathbf{R}/\mathbf{Z}$, i.e., the reals modulo 1. Correspondingly, a homeomorphism $f : S^1 \mapsto S^1$ has a *lift* $\tilde{f} : \mathbf{R} \mapsto \mathbf{R}$, i.e., a homeomorphism of $\mathbf{R}$ such that

$$f(x(\mathrm{mod}\,1)) = \tilde{f}(x) \qquad (\mathrm{mod}\,1).$$

If $\tilde{f}^*$ is another such lift, then $\tilde{f}^* - \tilde{f} = k \in \mathbf{Z}$.

The function $\tilde{f}$ is monotone and, according to whether it is increasing or decreasing, the homeomorphism f is *orientation preserving* or *orientation reversing*. We shall concentrate our attention on the former case (which is the one occurring in the Hopf bifurcation), and note that $\tilde{f}^* - \tilde{f} = k$ implies

$$\tilde{f}^{*n} - \tilde{f}^n = nk \tag{13.9}$$

For an orientation-preserving homeomorphism of S^1, we shall show that the following limit exists:

$$R(f) = \lim_{n \to \infty} \tfrac{1}{n} \tilde{f}^n x \tag{13.10}$$

and is independent of x, defining the *rotation number*[17] $R(f)$. If we interpret $R(f)$ as an element of $\mathbf{R}/\mathbf{Z}$, we see from (13.9) that it does not depend on the choice of the lift $\tilde{f}$. Here are some properties of the rotation number. $R(f)$ *is invariant under changes of coordinates on* S^1 [more precisely, $R(g^{-1}fg) = R(f)$ for every homeomorphism $g : S^1 \mapsto S^1$]; $R(f)$ *depends continuously on* f [if $g \to f$ uniformly on S^1, then $R(g) \to R(f)$]; *the homeomorphism* f *has a periodic point of period* q *if and only if* $R(f)$ *is rational of the form* p/q.

Let us now prove the existence of the limit (13.10). We write $\varphi_m(x) = \tilde{f}^m x - x$ and thus have $\varphi(x + \text{integer}) = \varphi(x)$. Notice also that if $0 \leqslant y - x < 1$, we have $0 \leqslant \tilde{f}^m y - \tilde{f}^m x < 1$, so that

$$|\varphi_m(y) - \varphi_m(x)| < 1. \tag{13.11}$$

Given $x \in \mathbf{R}$ and $m \in \mathbf{Z}$, we may choose $y \in \mathbf{R}$ such that $y = \tilde{f}^m x (\text{mod } 1)$, and $0 \leqslant y - x < 1$. Then

$$\varphi_{m+n}(x) = \varphi_m(x) + \varphi_n(y)$$
$$= \varphi_m(x) + \varphi_n(x) + (\varphi_n(y) - \varphi_n(x)).$$

Therefore, if $N = mq + r$, we have

$$|\varphi_N(x) - q\varphi_m(x) - \varphi_r(x)| < q. \tag{13.12}$$

Thus $\varphi_N(x) \approx q\varphi_m(x) + \varphi_r(x)$ for large m, and from this near-additivity property, one easily derives the existence of the limit

$$R(f) = \lim_{n \to \infty} \tfrac{1}{n} \varphi_n(x),$$

---

[17] The rotation number was first introduced by Poincaré.

which is furthermore independent of x in view of (13.11). The relation (13.10) immediately follows from this. One also easily proves the invariance of $R(f)$ under changes of coordinates, and the continuity of $f \mapsto R(f)$ [use (13.12)]. Suppose now that f has a periodic point x, of period q. We may then assume that $\tilde{f}^q x = x$, and therefore $qR(f) = R(f^q) = 0 \pmod 1$ so that $R(f)$ is rational of the form $p/q \pmod 1$. Conversely, if $R(f) = p/q$, let $g = f^q$. Since $R(g) = 0$, one can choose $\tilde{g}$ such that $\lim_{n \to \infty} \tilde{g}^n x / n = 0$. Therefore, $\tilde{g}$ has a fixed point x [otherwise $\inf |\tilde{g}x - x| > 0$ by compactness, contradicting $\lim_{n \to \infty} \tilde{g}^n x / n = 0$].

Having introduced rotation numbers for homeomorphisms of the circle, we return to the study of diffeomorphisms. Denote by $\mathrm{Diff}^r_+(S^1)$ the space of orientation preserving C^r diffeomorphisms of S^1, with $\mathbf{r} \geqslant 1$. We let Γ consist of those f such that $R(f)$ is rational and all the periodic points of f are hyperbolic.[18] One can show that Γ *is open and dense in* $\mathrm{Diff}^r_+(S^1)$ with respect to the C^r topology introduced in Section 2.7 (see Problem 7).

Consider now a family (f_μ) of diffeomorphisms of S^1, where μ varies over an open interval $J \subset \mathbf{R}$, and $(\mu, x) \mapsto f_\mu x$ is smooth. For instance, we may take

(13.13) $$f_\mu x = x + \mu \pmod 1.$$

In this example,

$$R(f_\mu) = \mu \pmod 1,$$

and f_μ is in fact the *rotation* by μ; in particular, f_μ is never in Γ, which is a rather unusual situation. Normally, for each rational number p/q, there is a whole open interval of μ's for which f_μ has a hyperbolic periodic orbit of period q, and $R(f_\mu) = p/q$. Furthermore, the set $\{\mu \in J : f_\mu \in \Gamma\}$ normally is dense in J, so that $R(f_\mu)$ is rational and locally constant on a dense open subset of J. We say that this situation is normal because it can be obtained by a small perturbation from any other situation. (A precise discussion would require a definition of *genericity* for families of diffeomorphisms of S^1.)

Since a dense open set is in some sense a large set, it would appear that $R(f_\mu)$ is rational for most μ's in the above setup. This, however, is not the case, because a dense open set in J may well have small Lebesgue

---

[18] Since S^1 is one-dimensional, a hyperbolic periodic orbit is either attracting or repelling. If $f \in \Gamma$, then f has *both* attracting and repelling periodic orbits.

measure compared to that of J. In fact, Arnold [1] has shown that for a small perturbation of the family (13.13), $R(f_\mu)$ remains irrational for most values of μ (in the sense of Lebesgue measure). He studies in particular the diffeomorphisms

$$(13.14) \qquad f_{\alpha,\mu}x = x + \alpha \sin 2\pi x + \mu \pmod 1$$

with $\alpha \in [0,1)$. The regions $\{(\alpha, \mu) : R(f_{\alpha,\mu}) = p/q\}$ are known as *Arnold's tongues*. Note that if we take $\alpha \geq 1$ in (13.14), $f_{\alpha,\mu}$ is no longer a diffeomorphism. Actually, for $\alpha > 1$, there appear zones of chaos (sensitive dependence on initial condition). This type of approach to chaos has been much studied (by *renormalization group* techniques) in the special case where the rotation number is fixed at the *golden ratio* value $\frac{1}{2}(\sqrt{5} - 1)$.[19]

If a diffeomorphism of the circle has an *irrational* rotation number, what does it look like? A classical theorem of Denjoy states that *if $f \in \mathrm{Diff}_+^{(1,1)}(S^1)$ and $R(f)$ is irrational, then f is topologically conjugate to the rotation by $R(f)$.* There is thus an orientation-preserving homeomorphism h of S^1 such that

$$hfh^{-1}(x) = xR(f) \pmod 1.$$

By a famous theorem of Herman, if f is sufficiently smooth and $R(f)$ *sufficiently irrational*, then f is differentiably conjugate to the rotation by $R(f)$. Yoccoz has shown that $R(f)$ is sufficiently irrational if it satisfies a *Diophantine condition* of order $\beta > 0$, i.e., if there is $C > 0$ such that $|R(f) - p/q| \geq C/q^{2+\beta}$ for every rational p/q. (The set of real numbers satisfying such a condition has full Lebesgue measure). Then, if $f \in \mathrm{Diff}_+^k(S^1)$ with integer $k \geq 3$, $2\beta + 1$, the conjugating homeomorphism h is in $\mathrm{Diff}_+^{\mathbf{r}}(S^1)$, where $|\mathbf{r}| > k - 1 - \beta - \varepsilon$ if $\varepsilon > 0$ (if $k = \infty$ or ω one can take $\mathbf{r} = \infty$, ω).[20]

Let us now return to the Hopf bifurcation for a map, and notice that the simple picture of a bifurcation separating two regions of structural

---

[19]The continued fraction $[1, 1, 1, \ldots]$ of $\frac{1}{2}(\sqrt{5} - 1)$ shows that this number is *most irrational*, i.e., the most difficult to approximate by rationals. For the renormalization group studies, see Feigenbaum, Kadanoff, and Schenker [1], Ostlund, Rand, Sethna, and Siggia [1], and Mestel [1] (the former papers are heuristic, the latter is a computer-aided proof).

[20]For Herman's theorem, see Herman [1], [2], and Yoccoz [1]; see also Khanin and Sinai [1].

stability (or Ω stability) fails here. If $R(f_\mu|H_\mu)$ is irrational, the nonwandering set for f_μ near 0 is $\{0\} \cup H_\mu$; if $f_\mu|H_\mu \in \Gamma$, the nonwandering set consists of 0 and the hyperbolic periodic points of $f|H_\mu$. Usually, there will thus be a sequence $\mu_n \to 0$ such that the only nonwandering points of f_{μ_n} near 0 are 0 and hyperbolic periodic points with period $q_n \to \infty$. Thus, a different dynamical behavior of f_μ occurs for infinitely many values of μ tending to 0.[21]

13.3. Theorem (Hopf for a semiflow). *Let U be an open subset of a Banach space E, J an interval of $\mathbf{R}$, and $(\mu, x, t) \mapsto f_\mu^t x$ a continuous map $J \times U \times [0, T] \mapsto E$ defining a local semiflow. We assume that $(\mu, x, t) \mapsto f_\mu^t x$ restricted to $J \times U \times (0, T)$ is of class C^2 and that, for fixed μ, $(x, t) \mapsto f_\mu^t x$ is $C^\mathbf{r}$ on $U \times (0, T)$ with $\mathbf{r} \geqslant 3$. We also assume that $(0, 0) \in J \times U$, that $f_\mu^t(0) = 0$, and that the spectrum of $D_0 f_\mu^t$ is contained (for $t \in (0, T)$) in $\{z : |z| < 1\}$ except for two eigenvalues α_μ^t, $\overline{\alpha}_\mu^t$ such that*

$$(a) \; |\alpha_0^t| = 1, \quad \tfrac{1}{2\pi i} \tfrac{d}{dt} \alpha_0^t|_{t=0} > 0,$$

$$(b) \; \tfrac{d}{d\mu} |a_\mu^t| > 0.$$

We write $f_\mu^\tau = f_\mu$ for $\tau \in (0, T)$, and make the generic assumption that the coefficient C computed as in Theorem 13.1 from derivatives up to order 3 of $f_0 = f_0^\tau$ does not vanish, and we restrict $J \times U$ to a suitable neighborhood of $(0, 0)$.

Under these conditions, the conclusions of Theorem 13.1 apply with the simple replacement of f_μ by f_μ^t. (Here, however, we have not restricted $\mathbf{r}$ to be $< \infty$). Furthermore, the set $\{(\mu, x) :$ the orbit of x under (f_μ^t) is periodic and contained in $U\}$ is the union of $J \times \{0\}$ and a two dimensional C^1 manifold H tangent at $(0, 0)$ to $\{0\} \times V$, where V is the subspace corresponding to the eigenvalues α_μ^t and $\overline{\alpha}_0^t$ of $D_0 f_0^t$. The manifold H consists of the point $(0, 0)$ and the periodic orbits $\{\mu\} \times H_\mu$. (Since H_μ is here a periodic orbit for (f_μ^t), all of its points are nonwandering). The period of H_μ tends to $T_0 = 2\pi i (\tfrac{d}{dt} \alpha_0^t|_{t=0})^{-1}$ when $\mu \to 0$.

A small change of τ will preserve the condition $C \neq 0$ and achieve $\alpha_0^3 \neq 1$, $\alpha_0^4 \neq 1$ (with $\alpha_0 = \alpha_0^\tau$). Therefore, Theorem 13.1 applies to

---

[21] For a detailed study of the sort of problems that arise here, we refer the reader to the extensive discussion by Chenciner [1].

$f_\mu = f_\mu^\tau$. If $0 < t < \tau$, the set $f_\mu^t H_\mu$ is again f_μ^τ invariant, and the property of uniqueness of H_μ in Theorem 13.1 yields $H_\mu \supset f_\mu^t H_\mu \supset f_\mu^t f_\mu^{\tau-t} H_\mu = H_\mu$. Therefore, $f_\mu^t H_\mu = H_\mu$ for all $t \geqslant 0$. Since α_0^t is not identically 1, H_μ contains no fixed point for (f_μ^t) when μ is small. This implies that H_μ is a periodic orbit: $H_\mu = \{f_\mu^t x\}$ (it is easy to check that the period tends to T_0 when $\mu \to 0$). In particular, H_μ is a C^r manifold.

It remains to show that the set

$$H = \{(0,0)\} \cup \bigcup_\mu (\{\mu\} \times H_\mu)$$

is a C^1 manifold tangent at $(0,0)$ to $\{0\} \times V$. Let $x_0 \in H_{\mu_0}$ and choose a piece of hyperplane Σ in E such that Σ is transversal to the orbit H_{μ_0} at x_0. The first return time $T(\mu, x)$ has a C^2 dependence on its arguments. The x-derivative of $f_\mu^{T(\mu,x)} x - x$ at (μ_0, x_0) is invertible $\Sigma \to \Sigma$ because of the normal hyperbolicity of H_{μ_0}. The Implicit Function Theorem B.3.3 thus implies that the trace of H in $J \times \Sigma$ is a C^2 manifold near (μ_0, x_0). Therefore, H is a C^2 manifold near (μ_0, x_0) (see Corollary B.3.5). A fortiori, H is a manifold of class C^1 except perhaps at the point $(0,0)$. But the proof of Theorem 13.1 shows that H is also differentiable at $(0,0)$ and that its tangent plane there is $\{0\} \times V$, concluding the proof.

13.4. Quasiperiodic Motions

We have just seen how a semiflow with an attracting fixed point could acquire an attracting periodic orbit through a Hopf bifurcation. A natural idea is that something like this could happen again and again, introducing more and more periods into the system. Indeed, the proposal was made by Hopf [2] that through successive bifurcations, one obtains *quasiperiodic flows* with an increasing number of periods.

To define a quasiperiodic flow, or quasiperiodic motion, with k periods, we need a k-dimensional torus T^k that is invariant for the semiflow (f^t). (We omit for the moment the subscript μ referring to the bifurcation parameter, and we think of T^k as an *attracting* manifold; see Section 14.) We say that $(f^t)|T^k$ is *quasiperiodic* if, for $x \in T^k$, we may write

$$f^t x = \Phi(\nu_1 t, \ldots, \nu_k t),$$

where Φ is periodic of period 1 separately in its k arguments, and the *frequencies* $\nu_1, \ldots, \nu_k$ are rationally independent, i.e., $\sum_1^k \ell_i \nu_i \neq 0$ when $\ell_1, \ldots, \ell_k \in \mathbf{Z}$ and $\sum_1^k |\ell_i| > 0$. (Note that the *periods* of the

quasiperiodic flow are related to the frequencies by $T_i = 1/\nu_i$.)
We shall assume that Φ is a smooth function; Φ parametrizes T^k by
k *angular variables* $\theta_1, \ldots, \theta_k$, and the time evolution (f^t) is linear in
terms of these: $\theta_i = \nu_i t$.

Can attracting tori $T^1, T^2, T^3, \ldots$ with quasiperiodic motions appear
through successive generic bifurcations? The first step, obtaining T^1
from an attracting fixed point, is just the Hopf bifurcation for a semiflow.
Note that the motion on T^1 is periodic, i.e., quasiperiodic with one
period. By use of a Poincaré section (see Fig. 21), the second step $T^1 \rightarrow$
T^2 reduces to the Hopf bifurcation for a map. If the eigenvalue α
occurring in Theorem 13.1 is not a root of unity, the motion on the
nascent T^2 is quasiperiodic with frequencies ν_1 and $\nu_2 = \nu_1 \cdot \operatorname{Im} \log \alpha$.
So far, so good. Unfortunately, further bifurcations to $T^3, T^4, \ldots$ do
not go so well. To understand this, we have to look into the stability of
quasiperiodic motions.

Let thus T^k be an attracting invariant torus for (f^t), and suppose
that (f^t) restricted to T^k is quasiperiodic. What will happen if (f^t) is
perturbed slightly and replaced by $(\tilde{f}^t)$? The torus T^k will be perturbed
slightly into a torus $\tilde{T}^k$. This is because of the persistence of normally
hyperbolic manifolds, to be discussed in Section 2.6: We use the fact
that (f^t) contracts towards T^k more strongly than it distorts distances
on T^k (note that the usual *flat* Riemann metric on T^k is preserved by
quasiperiodic motions).

We may identify $\tilde{T}^k$ and T^k (using a projection) and we now have to
analyze the effect of a small perturbation $(f^t) \mapsto (\tilde{f}^t)$ of a quasiperiodic flow on T^k. On T^1, the reader will check that we have stability:
A small perturbation of a periodic flow is again a periodic flow, with a
slightly different period. For T^2, by use of a Poincaré section, we can
reduce the discussion to that of diffeomorphisms of the circle T^1. As
we have seen in Section 13.2, a small perturbation can change an irrational rotation of T^1 into a rational rotation with hyperbolic periodic
points. On T^2 this corresponds to the replacement of a quasiperiodic
flow by a flow asymptotic to attracting periodic orbits (see Fig. 26).
This phenomenon is known as *frequency locking* because the ratio of
frequencies ν_2/ν_1, which was irrational for the quasiperiodic flow, *locks*
into a rational value p/q. There is an interesting situation, where ν_2/ν_1
is irrational, but close to a rational p/q, and the orbit $(f^t x)$ spends
most of its time following a *phantom* periodic orbit, from which it occa-
sionally unlocks. What is observed here is a nonstrange version of the

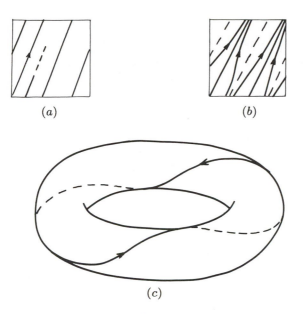

FIG. 26. Frequency-locking on T^2. The torus T^2 is pictured as a square with opposite sides identified. The quasiperiodic flow (a) gives by a small perturbation the flow (b) with one attracting and one repelling periodic orbit. ((c) is a perspective view of the torus, with the attracting periodic orbit.) The return map on the lower side of the squares (a), (b) is a Poincaré map; it is an irrational rotation in (a), but has periodic orbits of period 2 (one attracting and one repelling) in (b).

Manneville–Pomeau phenomenon discussed in Section 11.3. To summarize, the bifurcation $T^1 \to T^2$ in general leads to both quasiperiodic and periodic behavior alternating in a complicated manner.

Perturbation of a quasiperiodic flow on T^k for $k \geqslant 3$ leads to an even messier picture: Quasiperiodicity with k independent frequencies may persist,[22] or frequency locking may lead to periodic orbits or to quasiperiodic motion with less than k independent frequencies. Furthermore, *chaos* may also appear: One can show that *strange attractors* arise by (suitable) arbitrary small perturbations of any given quasiperiodic flow on T^k for $k \geqslant 3$.[23]

[22] See, in particular, the study of Broer, Huitema, and Takens [1].

[23] See Ruelle and Takens [1], Newhouse, Ruelle, and Takens [1]. These papers discuss Poincaré maps $T^{k-1} \mapsto T^{k-1}$ and show that in the spaces $C^2(T^2)$, or $C^\infty(T^{k-1})$ with $k \geqslant 4$, there are, arbitrarily close to any given irrational translation, diffeomorphisms with strange Axiom A attractors. (For a discussion of Axiom-A dynamical systems, see Appendix D.)

At this point, the reader who has access to a computer may want to look for himself or herself at the various types of behavior that arise by perturbation of a quasiperiodic flow on T^k, or equivalently by perturbation of an irrational translation on T^{k-1}. Taking $k = 3$, all one has to do is to program the map $f : T^2 \mapsto T^2$ such that

$$f(x,y) = (x + a + P(x,y) \pmod 1, \quad y + b + Q(x,y) \pmod 1),$$

where 1, a, b are linearly independent over the rationals, and P, Q are trigonometric functions of $2\pi x$, $2\pi y$. (Do not keep P, Q small, but try to keep f invertible!) A long orbit for f may show various types of behavior. Of particular interest is the situation where the points $f^n x$ accumulate on a curve $C \subset T^2$. If C is a smooth closed curve (i.e., a circle), $f|C$ is a diffeomorphism with (probably) an irrational rotation number. If C is folded repeatedly on itself, it is (probably) a strange attractor. Actually, the strange attractor may arise from a smooth circle when the normal contraction to this circle is no longer stronger than the internal distortion; an approximate model for this is given by a noninvertible map of a circle to itself (the renormalization group studies referred to in Section 13.2 are relevant to this approach to chaos).

Suppose then that a dynamical system tries to follow the scenario proposed by Hopf, of having a sequence of bifurcations producing tori T^k with increasing k, each torus carrying a quasiperiodic motion. What will happen generically is a very complicated set of bifurcations, where quasiperiodicity is seen part of the time, but also periodicity (due to frequency locking) and strange attractors. The production of strange attractors in this setting constitutes the quasiperiodic route to chaos, first proposed by Ruelle and Takens [1].

13.5. Turbulence and Chaos (A Digression)

Hopf [2] made his proposal of a sequence of quasiperiodic bifurcations in order to understand how hydrodynamic turbulence arises. In fact, Landau [1] made the same proposal independently (with somewhat less mathematical sophistication than Hopf). The analysis of Section 13.4 suggests that turbulence may not be quasiperiodic with many frequencies, but may correspond instead to the presence of strange attractors. This idea was proposed by Ruelle and Takens [1], and has later been experimentally confirmed.

We have at this point a reasonable understanding of the onset of turbulence: The *intermittent scenario of Manneville–Pomeau* (Section 11.3), the *Feigenbaum cascade* (Section 12.3), and the *quasiperiodic route of Ruelle–Takens* (Section 13.4) have all been observed.[24] This does not amount to a complete theory of turbulence (and we appear to be still quite far from understanding *fully developed turbulence*), but at least it is clear that sensitive dependence on initial condition must be a feature of any theory of hydrodynamic turbulence.

The reasons that suggest that strange attractors are present in hydrodynamics apply also to other dissipative physical systems. Conservative (i.e., Hamiltonian) systems cannot have attractors, but may exhibit sensitive dependence on initial condition. Indeed, chaos has now been seen in many natural systems (chemical, electromechanical, etc.). In some sense, this has been known for a long time. Hadamard [1] proved that there is sensitive dependence on the initial condition for the geodesic flow on a surface of constant negative curvature (a Hamiltonian system). Duhem [1] clearly understood the philosophical implications of Hadamard's result: Predictability of a deterministic system is limited by the unavoidable inaccuracies in the determination of initial conditions. Poincaré [4] discussed the relevance of this to weather prediction and the time evolution of a gas of hard spheres. These very modern ideas fell into oblivion for quite a while.

In more recent times, Lorenz [1], in a remarkable paper, studied a dynamical system in $\mathbf{R}^3$ numerically and showed that sensitive dependence on initial condition is present. The Lorenz attractor gives an approximate description of convection in a fluid layer. Since convection is present in the atmosphere, Lorenz could argue convincingly that sensitive dependence on initial condition precludes accurate long-term weather forecasts.

After the papers of Lorenz [1] and Ruelle and Takens [1], the interest in what is now called chaos developed first slowly, then explosively. Besides geometric differentiable dynamics and bifurcation theory, as discussed in the present monograph, the study of chaos uses the *ergodic theory of differentiable dynamical systems*,[25] which has come to play a prominent role. Unfortunately, differentiable dynamics—and particularly its ergodic theory—are difficult mathematical subjects, and much of the literature on chaos is vitiated by lack of understanding of basic mathematical facts. It is of course natural—and indeed desirable—that

---

[24] For further discussion of routes to turbulence, see Eckmann [1].

[25] For a review of the ergodic theory of differentiable dynamical systems as applied to chaos, see Eckmann and Ruelle [1].

there be some gap between rigorous mathematics and the methods used by students of natural phenomena. But the gap should not be too wide, and there is some danger that a subject that arose from close interplay of mathematics and physics would die from their divorce.[26]

14. Persistence of Normally Hyperbolic Manifolds

Consider a continuous family of smooth maps $f_\mu : M \mapsto M$, and let f_0 have a compact invariant manifold H_0 such that f_0 contracts or expands the directions transversal to H_0 more than it distorts H_0. Then, f_μ will have an invariant manifold H_μ for μ close to 0, and the smoothness class $C^\mathbf{r}$ of H_μ can be estimated.

14.1. Definition of r-Normal Hyperbolicity

Let (f^t) be a dynamical system such that each $f^t : M \mapsto M$ is $C^\mathbf{r}$, $1 \leqslant \mathbf{r} < \infty$. We assume that there is a compact C^1 submanifold H of M such that $f^t H = H$ and f^t restricted to H is a diffeomorphism for each t. (H is finite dimensional because it is compact, but M is a Banach manifold.) We assume that at each $x \in H$, there are subspaces V_x^-, V_x^+ of $T_x M$ such that $T_x H$, V_x^-, V_x^+ are complementary[27] in $T_x M$, and the corresponding projections depend continuously on x. This is expressed by saying that

(14.1) $$T_H M = TH + V^- + V^+$$

is a *continuous splitting* of the vector bundle $T_H M$ (see Appendix B.5 for the definition of vector bundles). We assume that the vector bundles $V^\pm$ are invariant in the sense that

$$(Tf^t)V_x^- \subset V_{fx}^-, \qquad (Tf^t)V_x^+ = V_{fx}^+.$$

Since $f^t|H$ is a diffeomorphism, we also have

$$(Tf^t)T_x H = T_{fx} H.$$

---

[26]The best introduction to a study of chaos is probably provided by the original articles on the subject. These are conveniently collected in Cvitanović [1] and Hao Bai-Lin [1], with surprisingly little overlap between the two.

[27]That is, $T_x H$, V_x^-, V_x^+ are linearly independent, and $T_x H + V_x^- + V_x^+ = T_x M$.

We denote by $T^0 f^t$, $T^{\pm} f^t$ the restrictions of Tf^t to $TH, V^{\pm}$. If we cover H by a finite number of charts of M, we obtain Banach norms on the spaces $T_x H$, $V_x^{\pm}$. Given a linear map $A : E \mapsto F$, we shall write

$$m(A) = \inf_{x:\|x\|=1} \|Ax\|$$

(see Appendix A.5).

We say that (f^t) is **r**-*normally hyperbolic* at H if one can choose $T > 0$ and $\theta > 1$ such that

(14.2)
$$\left.\begin{array}{c} \dfrac{\|T_x^- f^t\|}{m(T_x^0 f^t)^k} \leqslant \theta^{-t} \\[3ex] \dfrac{m(T_x^+ f^t)}{\|T_x^0 f^t\|^k} \geqslant \theta^t \end{array}\right\}$$

for all $k \in [0, |\mathbf{r}|]$ and $t \geqslant T$. (Remember that $|\mathbf{r}| = r + \alpha$ if $\mathbf{r} = (r, \alpha)$, and $|\mathbf{r}| = r$ if $\mathbf{r}$ is the integer r.) This definition does not depend on our choice of charts covering H (and, therefore, of norms on the $T_x H, V_x^{\pm}$) but the constants T, θ may depend on the choice.[28]

By taking $k = 1$ in (14.2), we see that the contractions or expansions in TH are exponentially small (for large t) compared with the contractions in V^- or expansions in V^+. In particular, the splitting (14.1) is unique. If $V^+ = 0$, H is said to be (**r**-normally) *attracting*.

In the following theorem, we discuss for simplicity the case of maps, but the results can be extended to semiflows (see Remark 14.3 (a)).

14.2. Theorem (Normally hyperbolic invariant manifolds). *Let H be a compact C^1 submanifold of M, and U an open neighborhood of H. We assume that $f : U \mapsto M$ is a $C^{\mathbf{r}}$ map, $1 \leqslant \mathbf{r} < \infty$, and that its restriction to H is a diffeomorphism of H. We also assume that f is* **r**-*normally hyperbolic at H, corresponding to the splitting*

$$T_H M = TH + V^- + V^+,$$

where the Banach spaces $V_x^{\pm}$ are separable and have the $C^{\mathbf{r}}$ extension property.

---

[28]If x is covered by several charts, there is an inconsequential ambiguity in the choice of norm on $T_x M$.

(Stable and unstable manifolds of H). *If U is replaced by a sufficiently small neighborhood of H, the stable and unstable manifolds*

$$\mathcal{V}_H^- = \bigcap_{n \geq 0} f^{-n} U,$$

$$\mathcal{V}_H^+ = \bigcap_{n \geq 0} f^{n} U$$

are C^r submanifolds of M, containing H and such that $T_x \mathcal{V}_H^\pm = T_x H + \mathcal{V}_x^\pm$ when $x \in H$ (in particular H is C^r).

(Stable and unstable manifolds of points of H). *Let a metric on U be defined by a distance d differing by a bounded factor from the norm distances corresponding to a finite family of charts of M covering U. Given $x \in H$, $m \geq 0$, denote by x_m the unique point of H such that $f^m x_m = x$. The stable and unstable manifolds*

$$\mathcal{V}_x^- = \{y \in \mathcal{V}_H^- : \lim_{n \to \infty} \|(T_x^0 f^n)^{-1}\| d(f^n y, f^n x) = 0\},$$

$$\mathcal{V}_x^+ = \{y \in \mathcal{V}_H^+ : (\exists y_m)_{m \geq 0}$$

$$\text{such that } y_0 = y, \quad f^n y_m = y_{m-n} \text{ for } n \leq m,$$

$$\text{and } \lim_{m \to \infty} \|T_{x_m}^0 f^m\| d(y_m, x_m) = 0\}$$

are C^r submanifolds of M, depending continuously on x for the C^r topology, such that $T_x \mathcal{V}_x^\pm = \mathcal{V}_x^\pm$ and $\bigcup_{x \in H} \mathcal{V}_x^\pm = \mathcal{V}_H^\pm$.[29]

(Persistence). *If $\tilde{f} : U \mapsto M$ is of class C^r and close to f for the C^r topology,[30] then $\tilde{f}$ is r-normally hyperbolic to a C^r manifold $\tilde{H}$ C^r close to H. Furthermore, the (un)stable manifolds of $\tilde{H}$, $x \in \tilde{H}$ for $\tilde{f}$ are C^r close to the corresponding manifolds for f.*

If we compare this theorem with the Center Stable and Center Unstable Manifold Theorem 7.1, we see that the normally hyperbolic invariant manifold H plays the role of center manifold, whereas $\mathcal{V}_H^-$ and $\mathcal{V}_H^+$ correspond to the center stable and center unstable manifolds. On the other hand, Theorem 14.2 may be viewed as an extension of the

---

[29] In the definition of $\mathcal{V}_x^\pm$, one could further impose that $d(f^n y, f^n x)$ or $d(y_m, x_m)$ tend to zero exponentially. In particular, $f^n y$ approaches H exponentially fast if $y \in \mathcal{V}_H^-$.

[30] Since U is covered by a finite family of charts of M, we can use the C^r topologies defined in Appendix B.1 for each chart. This defines what we call here C^r topology. Closeness of the (un)stable manifolds is similarly defined using charts and the C^r topologies.

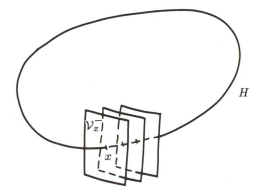

FIG. 27. The (strong) stable manifolds V_x^- for attracting compact manifold H.

Stable and Unstable Manifold Theorem 6.1, where the hyperbolic fixed point 0 is replaced by the hyperbolic invariant manifold H. Note that the manifolds $V_x^\pm$ are sometimes called *strong* (un)stable manifolds.

The manifolds $V_H^\pm$ and the families $(V_x^\pm)$ have local invariance properties that follow from their definitions. If H is attracting, V_H^- is a neighborhood of H, filled by the manifolds V_x^- (see Fig. 27).

14.3. Remarks

(a) **Semiflows.** The extension of Theorem 14.2 to a semiflow (f^t) presents no special difficulty if one assumes that $(x, t) \mapsto f^t x$ is continuous on $M \times [0, +\infty)$, and $(x, t) \mapsto T_x f^t$ continuous on $M \times (0, \infty)$.

(b) **Another formulation of normal hyperbolicity.** Instead of using $T_x H$, V_x^-, V_x^+, one can express r-normal hyperbolicity in terms of $T_x H$ and subspaces $\widetilde{V}_x^-$, $\widetilde{V}_x^+$ of the quotient Banach space $T_x M / T_x H$ (see Appendix A.4). It suffices to replace the operators $T_x^\pm f^t$ in (14.2) by the corresponding operators $\widetilde{V}_x^\pm \mapsto \widetilde{V}_{fx}^\pm$ (see Problem 2). In particular, the definition of an attracting invariant manifold can be given in terms of $T_H M$, TH, without knowledge of any invariant splitting.

(c) **Hyperbolic periodic orbit for a semiflow.** A hyperbolic periodic orbit Π for a semiflow (f^t) is a normally hyperbolic manifold for (f^t), or f^1, in view of the preceding remark. Therefore, the manifolds V_Π^-, V_Π^+ of Section 6.3 are, in fact, the stable and unstable manifolds of Theorem 14.2. Furthermore, stable and unstable manifolds V_x^-, V_x^+ are defined for all $x \in \Pi$ such that $V_\Pi^\pm = \bigcup_x V_x^\pm$.

(d) **Linearization of f.**[31] The Grobman–Hartman theorem extends to the present situation: If f is a diffeomorphism, then f is topologically conjugate to the restriction of Tf to the bundle $V^- + V^+$ near H; similarly, for a flow (f^t).

(e) **Continuity of foliations.** Although the family $(V_x^\pm)$ is continuous it is, in general, not a smooth foliation of $V_H^\pm$. Similarly, the dependence of $V_x^\pm$ on x, in general, is not smooth.

14.4. On the Proof of the Theorem on Normally Hyperbolic Manifolds

Normally hyperbolic manifolds have been studied using the method of graph transforms by Fenichel [1], and Hirsch, Pugh, and Shub [1]. Our definition of **r**-normal hyperbolicity corresponds to what is called eventual relative **r**-normal hyperbolicity by Hirsch, Pugh, and Shub.

It is technically harder to apply the graph transform method to construct $V_H^\pm$ here than when H is reduced to a point. The principle, however, is the same. When f is slightly perturbed, the manifolds $V_H^\pm$ change only slightly, and so does the intersection $V_H^- \cap V_H^+$. (For a hyperbolic fixed point, this persistence property follows from the implicit function theorem.) The study of $V_x^\pm$ can again be made by the graph transform method and is analogous to the study of the (un)stable manifold of a fixed point, but with a parameter $x \in H$ added.

The theorem on normally hyperbolic invariant manifolds given above can be proven by the techniques of Hirsch, Pugh, and Shub. This necessitates the assumption that the spaces $V_x^\pm$ are separable and have the $C^\mathbf{r}$ extension property (an assumption that may be too strong).[32] The reader who wishes to reconstruct a proof of Theorem 14.2 may find the following technical lemma useful (otherwise, this may be skipped).

14.5. Lemma. *There are C^1 charts $(U_\alpha, \varphi_\alpha, E_\alpha^0 \oplus E_\alpha^- \oplus E_\alpha^+)$ of M, covering the C^1 manifold H, such that $\varphi_\alpha(H \cap U_\alpha) = E_\alpha^0 \cap \varphi_\alpha(U_\alpha)$,*

---

[31] See Pugh and Shub [1] for a precise treatment.

[32] Our Theorem 14.2 corresponds to Theorem (4.1) in Hirsch, Pugh, and Shub [1], which assumes, however, that M is finite dimensional, f a diffeomorphism, and $\mathbf{r} = r$ an integer. These restrictions can easily be lifted, but since the proof involves C^0 approximation by smooth functions, we are led to assuming that $V_x^\pm$ are separable and have the $C^\mathbf{r}$ extension property (see Proposition B.4.2). The C^1 extension property is also used in the proof of Lemma 14.5, which introduces special coordinates, useful in the proof of the C^1 version of Theorem 14.2.

and $(T_x\varphi_\alpha)V_x^\pm = E_\alpha^\pm$ *for* $x \in H \cap U_\alpha$. *Furthermore, if* $U_\alpha \cap U_\beta \neq \phi$, *there is an invertible linear map* $I_{\beta\alpha} : E_\alpha^- \oplus E_\alpha^+ \mapsto E_\beta^- \oplus E_\beta^+$ *such that* $I_{\beta\alpha}E_\alpha^\pm = E_\beta^\pm$ *and*

$$\varphi_\alpha^{-1}(\varphi_\alpha x + y) = \varphi_\beta^{-1}(\varphi_\beta x + I_{\beta\alpha}y)$$

when $x \in U_\alpha \cap U_\beta$, $y \in E_\alpha^- \oplus E_\alpha^+$.

Since H is C^1, we may assume that at each $x_0 \in H$, there is a chart $(U, \varphi, E^0 \oplus E^- \oplus E^+)$ such that $\varphi(H \cap U) = E^0 \cap \varphi(U)$. By identifying U with φU and x_0 with 0, we may also assume that $V_0^\pm = E^\pm$, and we see that there are continuous maps $x \mapsto A^\pm(x)$ from a neighborhood of 0 in E^0 to linear operators

$$A^-(x) : E^- \mapsto E^0 \oplus E^+, \quad A^+(x) : E^+ \mapsto E^0 \oplus E^-$$

such that $A^\pm(0) = 0$ and $V_x^\pm$ is the graph of $A^\pm(x)$.

Since $E^\pm$ has the C^1 extension property (see Appendix B.1) there are C^1 functions $\varphi_n^\pm : E^\pm \mapsto \mathbf{R}$ with values in $[0, 1]$ such that $\varphi_n^\pm(y) \neq 0$ only if $\delta^n \geqslant \|y\| \geqslant \delta^{n+2}$, $\|D\varphi_n^\pm\| \leqslant C\delta^{-n}$, and $\sum_0^\infty \varphi_n^\pm(y) = 1$ for $\|y\| \leqslant \delta$ (in these formulae, $0 < \delta < 1$ and $C > 0$). By regularization we may also define sequences $(A_n^\pm)$ of C^1 functions $E^0 \mapsto \mathcal{L}(E^\pm, E^0 \oplus E^\mp)$ such that $\lim_{n\to\infty} A_n^\pm = A^\pm$ uniformly and $\|DA_n^\pm\| \leqslant \delta^{-n/2}$. The function $f : E^0 \oplus E^- \oplus E^+ \mapsto E^0 + E^- \oplus E^+$ defined by

$$f(x + y^- + y^+) = x + y^- + y^+ + \sum_{n=0}^\infty \varphi_n^-(y^-)A_n^-(x)y^-$$

$$+ \sum_{n=0}^\infty \varphi_n^+(y^+)A_n^+(x)y^+$$

is then C^1, $D_0 f = 1$, and $(D_x f)E^\pm = V_x^\pm$. Therefore, the map $\varphi = f^{-1}$ restricted to a small neighborhood $\widetilde{U}$ of O satisfies $\varphi(H \cap \widetilde{U}) = E^0 \cap \varphi(\widetilde{U})$, and $(T_x\varphi)V_x^\pm = E^\pm$ for $x \in H \cap \widetilde{U}$. By using a suitable partition of unity[33] (ω_α) on H it is now easy to define maps f^α with compatibility properties such that the $\varphi_\alpha = f_\alpha^{-1}$ verify the lemma.

---

[33] See Proposition B.4.2.

15. Hyperbolic Sets

This section introduces a generalization of the concept of hyperbolicity, important in later applications. We shall consider the case of C^r maps and C^r semiflows, $r \geq 1$. Specifically, we assume that $(x, t) \mapsto f^t x$ is continuous for $t \geq 0$, and C^r for $t > 0$.[34]

Let the compact set K be invariant under the map f (i.e., $fK = K$) and assume that f restricted to K is a homeomorphism. We say that K is a *hyperbolic set* for f is there is a continuous splitting $T_K M = V^- + V^+$ of the tangent bundle restricted to K such that

$$(Tf)V^- \subset V^-, \quad (Tf)V^+ = V^+ \qquad \text{(invariance)},$$

where $Tf|V^+$ is invertible and there are $C > 0$, $\theta > 1$ with

(15.1) $$\max_{x \in K} \|Tf^{\pm n}|V^\pm\| \leq C\theta^{-n} \qquad \text{for } n \geq 0.$$

The condition that $f|K$ be injective (i.e., a homeomorphism) is unnatural. One could do without it at the cost of using systematically the compact set

$$K^\dagger = \{(x_k) \in \prod_{k \leq 0} K : fx_{k-1} = x_k\}$$

and the homeomorphism $f^\dagger$ such that

$$f^\dagger(x_k) = (fx_k).$$

V^+ is now assumed to be a bundle over $K^\dagger$, but the definition (15.1) is otherwise unchanged. We may call K a *prehyperbolic set*, with *hyperbolic cover* $K^\dagger$. We shall discuss this situation in Section 15.6 but here we proceed, for simplicity, with the assumption that $f|K$ is a homeomorphism.

Suppose now that the compact set K is invariant under the semiflow (f^t) and that K contains no fixed point. We say that K is a *hyperbolic set* for (f^t) if there is a continuous invariant splitting

$$T_K M = V^0 + V^- + V^+,$$

---

[34]We shall later need a topology on these dynamical systems. Since a neighborhood of the compact set K is covered by a finite family of charts $(U_\alpha, \varphi_\alpha, E_\alpha)$ of M, we can use $C^r(\varphi_\alpha U_\alpha, \psi_\beta V_\beta)$ topologies (see Appendix B.1) to define a "local" C^r topology for maps f that sent K close to itself. For a semiflow, we use the topology of uniform convergence of $(x, t) \mapsto f^t x$ on the $U_\alpha \times [0, 1]$, and the C^r topologies on the $U_\alpha \times (a, b)$, $0 < a < b < 1$.

where V^0 is the one-dimensional bundle in the direction of the flow ($V^0 = \mathbf{R}X$, where $X = \frac{d}{dt} f^t x$ at $t = 0$)[35] and there are $C > 0$, $\theta > 1$ with

(15.2) $$\max_{x \in K} \|Tf^{\mp t}|V^{\pm}\| \leqslant C\theta^{-t} \quad \text{for} \quad t \geqslant 0.$$

The set K is also allowed to contain isolated fixed points $a_1, \ldots, a_N$; we impose at each of them the hyperbolicity condition of Section 5.3.

We may use a finite number of charts covering K to define the norms in (15.1) or (15.2). Equivalently, we may assume that a norm is defined on each $T_x M$, with continuous dependence on $x \in K$. (If M is a finite-dimensional or Hilbert manifold, use a Riemann metric; see Appendix B.7). If K is a fixed point or periodic orbit, the above definitions reduce to those of Section 5. Compare also with the definition of normal hyperbolicity in Section 14.1.

Note that the maps $x \mapsto V_x^{\pm}$ are in general not differentiable, even if f is very smooth. Under mild assumptions, one can prove that these maps are Hölder continuous in the case of a hyperbolic set for a map (see Problem 4).

15.1. A Hyperbolicity Criterion[36]

Let f be a $C^{\mathbf{r}}$ map ($\mathbf{r} \geqslant 1$) and K a compact set such that $fK = K$ and $f|K$ is a homeomorphism. Given a splitting $T_K M = E^- + E^+$ (in general not invariant) and functions $\rho^{\pm} \geqslant 0$ on K, we define *sectors* $S^{\pm}$ by

$$S^-(x) = \{\xi^- + \xi^+ : \xi^{\pm} \in E_x^{\pm} \quad \text{and} \quad \|\xi^+\| \leqslant \rho^-(x)\|\xi^-\|\},$$
$$S^+(x) = \{\xi^- + \xi^+ : \xi^{\pm} \in E_x^{\pm} \quad \text{and} \quad \|\xi^-\| \leqslant \rho^+(x)\|\xi^+\|\}.$$

The hyperbolicity of K is already implied by the following condition.

(**Approximate hyperbolicity**). *There are $\lambda > 1$ and a positive integer m such that*

$$Tf^m S^+ \subset S^+, \qquad Tf^{-m} S^- \subset S^-$$
$$Tf^m E^+ + E^- = E$$

and

$$\|Tf^m \xi\| \geqslant \lambda\|\xi\| \qquad \text{if} \quad \xi \in S^+$$
$$\|Tf^m \xi\| \leqslant \lambda^{-1}\|\xi\| \qquad \text{if} \quad \xi \in S^-.$$

---

[35] This makes sense, because if $x \in K$, one may write $x = f^\tau y$ with $\tau > 0$.

If this condition is satisfied, a hyperbolic splitting of $T_K M$ is defined by $V^{\pm} = \lim_{k\to\infty} T f^{\pm km} S^{\pm}$ (see Problem 3). It is not necessary to assume that the splitting $E^- + E^+$ or the functions $\rho^{\pm}$ are continuous.

A similar criterion could be obtained for flows by discussing the quotient bundle maps $T_K M/V^0 \mapsto T_K M/V^0$ associated with the $T f^t$.

15.2. Stable and Unstable Manifolds. Persistence of Hyperbolic Sets

We first discuss the case of a C^r map f, with $r \geq 1$. If K is a compact hyperbolic set for f, a local stable manifold V_x^- and a local unstable manifold V_x^+ are defined for each $x \in K$ by

$$V_x^- = \{y \in M : d(f^n y, f^n x) < R \text{ for } n \geq 0\},$$
$$V_x^+ = \{y_0 \in M : \exists (y_k)_{k \leq 0} \text{ with } f y_{k-1} = y_k,$$
$$\text{and } d(y_k, f^k x) < R \text{ for } k \leq 0\}$$

In these formulae the metric d is defined on a neighborhood of K covered by a finite number of charts. We assume that d differs by a bounded factor from the norm distances defined by the charts,[37] and that R is a suitably small number. The following result then holds.

(Hyperbolic sets for maps: stable and unstable manifolds, persistence).[38] *The manifolds $V_x^{\pm}$ are of class C^r, and tangent at x to $V_x^{\pm}$. There are constants $C' > 0$, $\theta' > 1$ such that, if $y, z \in V_x^{\pm}$,*

$$d(f^{\mp n} y, f^{\mp n} z) \leq C' \theta'^{-n} d(y, z) \qquad \text{for } n \geq 0$$

(we choose $f^{-n} y$, $f^{-n} z$ inductively on n such that these points belong to $V_{f^{-n} x}^+$). Furthermore the map $x \mapsto V_x^{\pm}$ is continuous $K \mapsto C^r$.

If $\tilde{f}$ is C^1 close to f, there is a unique map $h : K \mapsto M$ close[39] to the canonical injection $i : K \hookrightarrow M$ and such that $\tilde{f} \circ h = h \circ f$ on K. If $\tilde{f}$ is injective on $\tilde{K} = hK$, then h is a homeomorphism and $\tilde{K}$ is a hyperbolic set for $\tilde{f}$. The stable and unstable manifolds of $\tilde{K}$ depend continuously on $\tilde{f}$ for the C^r topologies.

Let us make more precise the sense in which $V_x^{\pm}$ depends continuously on x (the dependence on f is handled similarly). Choose a chart

---

[37] The existence of such a metric may be proved along the lines of Appendix B.7.

[38] See Hirsch and Pugh [1].

[39] Closeness is with respect to the uniform distance, i.e., $\max_{x \in K} d(h(x), x)$ is small.

$(U, \varphi, E^+ \oplus E^-)$ at $a \in K$, with $E^\pm = V_a^\pm$. For x close to a, the set $\varphi V_x^\pm$ is the graph of a function from $D_x \subset E^\pm$ to $E^\mp$. One may choose $\varepsilon > 0$ so that D_x contains the ball $B^\pm$ of radius ε centered at $\varphi^\pm(x)$ in $E^\pm$. Thus, $\varphi V_x^\pm$ defines a C^r function $g_x : B^\pm \mapsto E^\mp$, or by translation a function $h_x : E_0^\pm(\varepsilon) \mapsto E^\mp$. Our continuity assertion is that the map $x \mapsto h_x$ is continuous with respect to the C^r topology on functions $E_0^\pm(\varepsilon) \mapsto E^\mp$ (see Appendix B.1).

Consider now a C^r semiflow (f^t), with $r \geqslant 1$. Let K be a compact hyperbolic set for (f^t) containing no fixed point. If $x \in K$, we define

$$V_x^{0-} = \{y \in M : d(f^t x, f^t y) < R \text{ for all } t \geqslant 0\},$$
$$V_x^- = \{y \in V_x^{0-} : \lim_{t \to \infty} d(f^t x, f^t y) = 0\},$$
$$V_x^{0+} = \{y_0 \in M : \exists (y_t)_{t \leqslant 0}, \quad d(f^{-s} x, y_{-s}) < R$$
$$\text{and} \quad f^u y_{-s} = y_{u-s} \text{ for } 0 \leqslant u \leqslant s\},$$
$$V_x^+ = \{y_0 \in V_x^{0+} : \lim_{s \to \infty} d(f^{-s} x, y_{-s}) = 0 \text{ with } (y_{-s}) \text{ as above}\}.$$

The $V_x^\pm$ are local (strong) stable and unstable manifolds, the $V_x^{0\pm}$ are local (center) stable and unstable manifolds.

(Hyperbolic sets for semiflows: stable and unstable manifolds, persistence). *The manifolds $V_x^\pm$, resp. $V_x^{0\pm}$, are of class C^r and tangent at x to $V^\pm$, resp. $V^0 + V^\pm$. Locally (i.e., close to x), $V_x^{0\pm}$ is a union of $f^t V_x^\pm$ for small $|t|$. There are constants $C' > 0$, $\theta' > 1$ such that, if $y, z \in V_x^\pm$,*

$$d(f^{\mp t} y, f^{\mp t} z) \leqslant C' \theta'^{-t} d(y, z) \qquad \text{for} \qquad t \geqslant 0.$$

Furthermore, the maps $x \mapsto V_x^\pm, V_x^{0\pm}$ are continuous $K \mapsto C^r$.

If $(\tilde{f}^t)$ is C^1 close to (f^t) there is a map $h : K \mapsto M$ close to the canonical injection $i : K \hookrightarrow M$ and mapping the orbits of $(f^t|K)$ to orbits of $(\tilde{f}^t)$. The map h is unique up to small displacements along the orbits. If the maps $\tilde{f}^t$ are injective on $\tilde{K} = hK$, then h can be taken to be a homeomorphism, and $\tilde{K}$ is a hyperbolic set for $(\tilde{f}^t)$. The stable and unstable manifolds of $\tilde{K}$ depend continuously on $(\tilde{f}^t)$ for the C^r topologies.

For semiflows (or flows), the map h cannot, in general, be chosen have $\tilde{f}^t h = h f^t$, and h is not a topological conjugacy). Also, h is not uniquely determined, since composition with orbit preserving homeomorphisms close to the identity is allowed.

Notice that in the case of diffeomorphisms and flows, one can define global stable and unstable manifolds:

$$\mathcal{W}_x^\pm = \bigcup_t f^t \mathcal{V}_{f^{-t}x}^\pm.$$

To define the local stable and unstable manifolds, it suffices to have a local map or semiflow (near K).

Now a few words on proofs. The manifolds $\mathcal{V}_x^\pm$ for a map can be studied by the graph transform method. In fact, if one considers the Banach manifold $\mathcal{M}$ of bounded maps $K \mapsto M$ (not necessarily continuous) and defines $\mathcal{F}: \mathcal{M} \mapsto \mathcal{M}$ by $\mathcal{F}h = f \circ h \circ (f|K)^{-1}$, it is found that the canonical inclusion $i : K \hookrightarrow M$ is a hyperbolic fixed point of $\mathcal{F}$.[40] The points of the (un)stable manifold of i are maps $k : K \mapsto M$, and the points $k(x)$ constitute the (un)stable manifold of $x \in K$. This elegantly reduces the study of a hyperbolic set to that of a hyperbolic fixed point. To prove persistence, let $\widetilde{\mathcal{F}}h = \tilde{f} \circ h \circ (f|K)^{-1}$. For $\tilde{f}$ close to f, $\widetilde{\mathcal{F}}$ has a unique fixed point h close to i, again hyperbolic, such that $\tilde{f} \circ h = h \circ f$ on K.[41]

Unfortunately, there is no simple reduction of the study of a hyperbolic set for a semiflow (or of a normally hyperbolic manifold, Section 14) to the study of a hyperbolic fixed point. One may again consider the manifold $\mathcal{M}$ of bounded maps $K \mapsto M$, and define $\mathcal{F}^t h = f^t \circ h \circ (f^t|K)^{-1}$. The inclusion $i : K \hookrightarrow M$ is a fixed point for $(\mathcal{F}^t)$ but it is not hyperbolic (think of the many maps that are just small displacements of K along its orbits). Nevertheless, i has strong stable and unstable manifolds and those permit the definition of the $\mathcal{V}_x^\pm$. Locally one can then define $\mathcal{V}_x^{0\pm} = \bigcup_{|t| \leq \tau} f^t \mathcal{V}_x^\pm$ and prove the asserted properties of invariant manifolds.[42] The method, however, does not yield the persistence under perturbations. To deal with this problem,

---

[40] If M is a Banach space, let E be the Banach space of all bounded maps $e : K \mapsto M$ with the norm

$$\|e\| = \max_{x \in K} |e(x)|.$$

The applications $K \mapsto M$ close to i correspond to maps $h : x \mapsto x + e(x)$, with small e. We may thus identify the manifold $\mathcal{M}$ with the chart E. A tangent vector $\in T_i \mathcal{M}$ is a section of $T_K M$, i.e., a bounded map $x \mapsto T_x M$ for $x \in K$. Hyperbolicity of i for $\mathcal{F}$ corresponds to the fact that K is a hyperbolic set for f.

[41] See Hirsch and Pugh [1]. The extension to a Banach manifold M and noninteger $\mathbf{r}$ are easy.

[42] See Hirsch, Palis, Pugh, and Shub [1].

one can use the fact that the maps f^τ are normally hyperbolic to the orbits $\{f^t x\}$ contained in K. These orbits are not compact, but K is, and one can use the graph transform method to establish the desired result.[43]

15.3. Expansiveness

The map f restricted to the hyperbolic invariant set K is an *expansive homeomorphism*. This means that there exists $\varepsilon > 0$ (*expansive constant*) such that $d(f^k x, f^k y) \leq \varepsilon$ for all $k \in \mathbf{Z}$ implies $x = y$. Similarly, the semiflow (f^t) restricted to the hyperbolic invariant set K is an *expansive flow*. This means that there exists $\varepsilon > 0$ (*expansive constant*) such that if there is a continuous function $s : \mathbf{R} \mapsto \mathbf{R}$ with $s(0) = 0$ and $d(f^t x, f^{s(t)} y) \leq \varepsilon$ for all $t \in \mathbf{R}$, then $y = f^\tau x$ (and $|\tau|$ is small for small ε).

To prove expansiveness in the map case, we notice that, when $\varepsilon < R$ (R small as before), the condition $d(f^k x, f^k y) \leq \varepsilon$ for all k implies $y \in V_x^- \cap V_x^+$. Since the local stable and unstable manifolds depend continuously on x, and since K is compact, we can choose ε such that the $V_x^\pm$ are *nearly flat*, i.e., uniformly C^1 close to (small) pieces of $V_x^\pm$. It follows readily that V_x^- and V_x^+ intersect transversally and that $V_x^- \cap V_x^+ = \{x\}$.

In the semiflow case, we find similarly that $y \in V_x^{0-} \cap V_x^{0+} \subset \{f^\tau x : |\tau| \text{ small}\}$.

15.4. Local Product Structure

Let K be a hyperbolic set for a map or semiflow. We say that K has *local product structure* if R can be chosen in the definition (Section 15.2) of $V_x^\pm$ such that, for all $x, y \in K$,

$$V_x^- \cap V_y^+ \subset K.$$

We may also assume that the $V_x^\pm$ are nearly flat so that $V_x^- \cap V_x^+$ consists of at most one point.

In the case of a map, there is an $\varepsilon > 0$ such that $V_x^- \cap V_y^+$ consists of exactly one point $[x, y]$ when $d(x, y) < 2\varepsilon$. Furthermore, there is $L > 0$ such that

(15.3)
$$\left. \begin{array}{l} d(x, [x, y]) \leq L d(x, y) \\ d(y, [x, y]) \leq L d(x, y) \end{array} \right\}.$$

---

[43] See Hirsch, Pugh, and Shub [1].

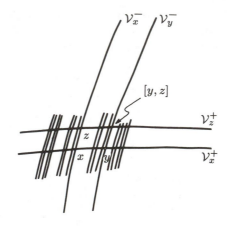

FIG. 28.   Local product structure (map case). To a pair $(y, z) \in \mathcal{V}_x^+ \times \mathcal{V}_x^-$ with $y, z \in K$, $d(x, y) \leq \varepsilon$, $d(x, z) \leq \varepsilon$, there corresponds a unique point $[y, z] \in K$. We have drawn $\mathcal{V}_x^+ \cap K$ as a Cantor set and $\mathcal{V}_x^- \cap K$ as a curve.

We now justify the name of the local product structure. Define

$$K_x^{\pm} = \mathcal{V}_x^{\pm} \cap \overline{B}_x(\varepsilon) \cap K$$

for some small $\varepsilon > 0$. In the map case, there is a homeomorphism of the product $K_x^+ \times K_x^-$ to a neighborhood of x in K, namely,

$$(y, z) \mapsto [y, z]$$

(see Fig. 28). In the semiflow case, there is, for small $\delta > 0$, a homeomorphism of $K_x^+ \times K_x^- \times [-\delta, +\delta]$ to a neighborhood of x in K given either by

$$(y, z, t) \mapsto [y, z, t] \in \mathcal{V}_{f^t y}^- \cap \mathcal{V}_z^{0+}$$

(see Fig. 29) or (inequivalently!) by

$$[y, z, t] \in \mathcal{V}_y^{0-} \cap \mathcal{V}_{f^t z}^+.$$

These facts result from the transversal intersection of the nearly flat manifolds involved.

15.5. Shadowing for Maps

A remarkable feature of hyperbolic sets with local product structure is that δ-pseudoorbits (see Section 8.3) are well approximated by true

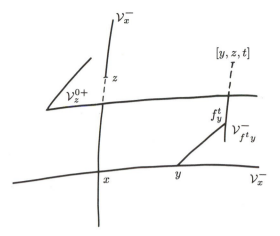

FIG. 29. Local product structure (semi-flow case). There is a homeomorphism of a neighborhood of x in K to the product of a piece of $\mathcal{V}_x^0 \cap K$ and a piece of $\mathcal{V}_x^- \cap K$.

orbits. To make this statement precise, we take $\varepsilon > 0$ and introduce the notion of ε-shadowing of a δ-pseudoorbit $(x_t)_{t \in [t_0, t_1]}$ by an orbit $(f^t x)$.

Remember that $(x_t)_{t \in [t_0, t_1]}$ is a δ-pseudoorbit for the map f if $d(fx_t, x_{t+1}) < \delta$ for every finite $t \in [t_0, t_1 - 1]$, where t_0, t_1 may be finite or infinite. The pseudoorbit (x_t) is ε-shadowed by the orbit $(f^t x)$ if $d(f^t x, x_t) < \varepsilon$ for all $t \in [t_0, t_1]$.

(Shadowing Lemma).[44] *Let K be a hyperbolic set with local product structure for the map f. For every $\varepsilon > 0$, there is a $\delta > 0$ such that every δ-pseudoorbit in K is ε-shadowed by an orbit in K.*

We indicate how the proof works when t_0 and t_1 are finite (otherwise, take limits $t_0 \to -\infty$ and/or $t_1 \to \infty$). The idea is to consider orbits $(y_t^{(\ell)})$ that approximate (x_t) on progressively longer intervals $[\ell, t_1]$ (a longer interval means a smaller ℓ). The orbit $(y_t^{(\ell)})$ is determined by induction on ℓ so that

$$y_\ell^{(\ell)} = x_{t_1} \qquad\qquad \text{when} \quad \ell = t_1$$
$$y_\ell^{(\ell-1)} = [y_\ell^{(\ell)}, fx_{\ell-1}] \quad \text{when} \quad \ell \leqslant t_1$$

(see Fig. 30), and of course $f^t y_s^{(\ell)} = y_{s+t}^{(\ell)}$.

[44] See Bowen [5], Newhouse [3].

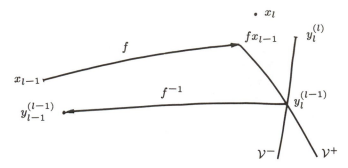

FIG. 30.   Determination of an orbit which ε-shadows a δ-psuedoorbit. The orbit $(y_t^{(\ell)})_{t\in[\ell,t_1]}$ is defined inductively for decreasing ℓ until $\ell = t_0$.

In view of (15.3), we have

$$d(fx_{\ell-1}, y_\ell^{(\ell-1)}) \leqslant Ld(fx_{\ell-1}, y_\ell^{(\ell)}).$$

We shall suppose that—for some $\alpha < 1$—the map f contracts the distances in stable directions by a factor $< \alpha$ and expands the distances in unstable directions by a factor $> L/\alpha$. (To achieve this it may be necessary to replace f by some iterate f^N.) In particular,

$$d(x_{\ell-1}, y_{\ell-1}^{(\ell-1)}) \leqslant \tfrac{\alpha}{L}d(fx_{\ell-1}, y_\ell^{(\ell-1)}) \leqslant \alpha d(fx_{\ell-1}, y_\ell^{(\ell)})$$
$$\leqslant \alpha[\delta + d(x_\ell, y_\ell^{(\ell)})].$$

If δ is so small that $\alpha(\delta + \varepsilon_0) < \varepsilon_0$, we have $d(x_{\ell-1}, y_{\ell-1}^{(\ell-1)}) < \varepsilon_0$ by induction, so that $d(x_\ell, y_\ell^{(\ell)}) < \varepsilon_0$ for all ℓ. Furthermore, again using (15.3),

$$d(y_\ell^{(\ell)}, y_\ell^{(\ell-1)}) \leqslant Ld(fx_{\ell-1}, y_\ell^{(\ell)}) \leqslant L(\delta + \varepsilon_0),$$

and therefore,

$$d(x_t, y_t^{(\ell)}) \leqslant d(x_t, y_t^{(t)}) + \sum_{k=\ell+1}^{t} d(f^{t-k}y_k^{(k)}, f^{t-k}y_k^{(k-1)})$$
$$\leqslant \varepsilon_0 + L(\varepsilon_0 + \delta)(1 + \alpha + \alpha^2 + \cdots)$$
$$< (\varepsilon_0 + \delta)(1 + L)/(1 - \alpha).$$

This is $< \varepsilon$ if δ is sufficiently small.

(Existence of a fundamental neighborhood).[45] *If K has a local product structure, there is a neighborhood U of K (fundamental neighborhood) such that if $f^n y \in U$ for $n \geqslant 0$ (resp., there are $y_k \in U$ with $y_0 = y$ and $f y_{k-1} = y_k$ for $k \leqslant 0$) then $y \in V_x^-$ (resp. $y \in V_x^+$) for some $x \in K$. Furthermore,*

$$(15.4) \qquad \bigcap_{k \in \mathbf{Z}} (f|U)^k U = K.$$

If $f^k y \in U$ for $k \geqslant 0$, and U is a small neighborhood of K, we may choose $y_k \in K$ close to $f^k y$ for each k, and it follows that (y_k) is a δ-pseudoorbit with small $\delta > 0$. Shadowing gives an orbit $(f^k x)$, where $x \in K$ and $d(f^k x, f^k y) < \varepsilon$ for all $k \geqslant 0$, hence $y \in V_x^-$. Similarly for V_x^+. If $(f|U)^k$ is defined on y for all $k \in \mathbf{Z}$, we find $y \in V_x^- \cap V_{x'}^+ \subset K$ by the local product structure, and (15.4) follows.

(Persistence). *If K has local product structure, there is a fundamental neighborhood U of K and a neighborhood $\mathcal{N}$ of f in the C^1 maps such that if $\tilde{f} \in \mathcal{N}$, then $\bigcap_{k \in \mathbf{Z}} (\tilde{f}|U)^k U$ is the set $\tilde{K} = hK$ associated with $\tilde{f}$ in Section 15.2.*

We first assume that $\tilde{f}|\tilde{K}$ is a homeomorphism, and sketch the proof in that case. Since $\tilde{f}$ is close to f, the construction of the (un)stable manifolds $\tilde{V}^\pm$ of $\tilde{K}$ shows that (uniformly in $\tilde{f} \in \mathcal{N}$) they are close to the (un)stable manifolds of K. Suppose we know that $d(y_k, \tilde{f}^k u) < 2\varepsilon$ and $\tilde{f} y_k = y_{k+1}$, all $k \in \mathbf{Z}$, with ε suitably small and $u \in \tilde{K}$; then we have $y_0 \in \tilde{V}_u^- \cap \tilde{V}_u^+ = \{u\}$. If we only know that $y_k \in U$ for all k, the condition $\tilde{f} y_k = y_{k+1}$ implies that (y_k) is a δ-pseudoorbit for f, hence is ε-shadowed by $(f^k x)$ with $x \in K$, hence is 2ε-shadowed by $(\tilde{f}^k u)$ with $u = hx$, and we may thus conclude that $y_0 = hx$.

If $\tilde{f}|\tilde{K}$ is not injective, $\tilde{K}$ is only a prehyperbolic set, and we have to use, instead of $u \in \tilde{K}$, a point $(hf^k x)_{k \leqslant 0}$ of its hyperbolic cover $\tilde{K}^\dagger$, but the result is the same.

15.6. Prehyperbolic Sets

As we see at this point, the condition that $f|K$ be a homeomorphism is unnatural. Let us now omit this condition, i.e., assume only that K is

---

[45] See Hirsch, Palis, Pugh, and Shub [1]. The proof in that paper does not use shadowing.

prehyperbolic. We define the *hyperbolic cover*

$$K^\dagger = \{(x_k) \in \prod_{k \leqslant 0} K : fx_{k-1} = x_k\}$$

and the homeomorphism $f^\dagger : K^\dagger \mapsto K^\dagger$ such that

$$f^\dagger(x_k) = (fx_k).$$

(Notice that $K^\dagger$ is homeomorphic to the space of orbits $(x_k)_{k \in \mathbf{Z}}$, with $f^\dagger$ acting as the shift $(x_k) \mapsto (x_{k+1})$.) We may introduce a metric d_α on $K^\dagger$ by

$$d_\alpha((x_k), (y_k)) = \sup_{k \leqslant 0} \alpha^{|k|} d(x_k, y_k)$$

for any $\alpha \in (0, 1)$ and write

$$\pi((x_k)) = x_0.$$

The map $\pi : K^\dagger \mapsto M$ projects $K^\dagger$ onto K, contracts distances, and satisfies

$$\pi \circ f^\dagger = f \circ \pi.$$

With these definitions, we can replace $\mathcal{M}$ in Section 15.2 by the manifold $\mathcal{M}^\dagger$ of maps $h : K^\dagger \mapsto M$, and write $(\mathcal{F}h)((x_k)) = (f \circ h)((x_{k-1}))$. The map π is a hyperbolic fixed point for $\mathcal{F}$. If $\tilde{f}$ is a C^1 small perturbation of f, we thus find $h : K^\dagger \mapsto M$ such that

$$(\tilde{f} \circ h)((x_{k-1})) = h((x_k)).$$

The set $\widetilde{K} = hK^\dagger$ is prehyperbolic, and h lifts to a homeomorphism $h^\dagger : K^\dagger \mapsto \widetilde{K}^\dagger$.

We define stable and unstable manifolds:

$$\mathcal{V}_x^- = \{y \in M : d(f^n y, f^n x) < R \quad \text{for } n \geqslant 0\}$$
$$\mathcal{V}_{(x_k)}^+ = \{y_0 \in M : \exists (y_k)_{k \leqslant 0} \quad \text{with } fy_{k-1} = y_k$$
$$\text{and } d(y_k, x_k) < R \quad \text{for } k \leqslant 0\}$$

for $x \in K$ and $(x_k) \in K^\dagger$. (We assume that R is small.) We say that $K^\dagger$ has *local product structure* if

$$\mathcal{V}_x^- \cap \mathcal{V}_{(y_k)}^+ \subset K.$$

Since $\mathcal{V}^\pm$ may be assumed nearly flat, the intersection consists of one point at most. Thus there is ε such that if $d(x, y_0) < 2\varepsilon$, then $\mathcal{V}_x^- \cap \mathcal{V}_{(y_k)}^+$ consists of exactly one point $[x, (y_k)]$. If $K^\dagger$ has local product structure, we may again prove that K has a fundamental neighborhood U, such that if an orbit remains in U for all n, it is in K. If $\tilde{f}$ is C^1 close to f, then

$$\widetilde{K} = \{y_0 : (y_k)_{k \in \mathbf{Z}} \text{ is an } \tilde{f} \text{ orbit contained in } U\}.$$

Of course, if f is injective, the prehyperbolic set K is hyperbolic. If f is a diffeomorphism, the same holds for the perturbed set $\widetilde{K}$.

Incidentally, if K is hyperbolic (resp. has local product structure) for a diffeomorphism f, then K is hyperbolic (resp. has local product structure) for f^{-1}.

15.7. Shadowing for Flows

The theory of shadowing for semiflows is similar to that of shadowing for maps. Here, however, it is necessary to change the time parametrization in order to match an orbit and a pseudoorbit. (Unit intervals of time $(k, k + 1)$ for the orbit are matched with intervals of time of length $u_k \approx 1$ for the pseudoorbit.) Because of this complication, we refrain from an explicit formulation but note some consequences valid for a set K with local product structure for a semiflow (f^t).

(**Existence of a fundamental neighborhood**).[46] *If K has local product structure, there is a neighborhood U of K (fundamental neighborhood) such that if $f^t y \in U$ for $t \geq 0$ (resp. there are $y_s \in U$, $s \leq 0$ with $y_0 = y$ and $f^t y_s = y_{t+s}$ if $t + s \leq 0$), then $y \in \mathcal{V}_x^-$ (resp. $\mathcal{V}_x^+$) for some $x \in K$. Furthermore, if $(y_s)_{s \in \mathbf{R}}$ is an orbit contained in U, then $y_0 \in K$.*

(**Persistence**).[47] *If K has local product structure, there is a fundamental neighborhood U of K and a neighborhood $\mathcal{N}$ of (f^t) in the C^1 semiflows such that if $(\tilde{f}^t) \in \mathcal{N}$, then $\{y_0 : (y_s)_{s \in \mathbf{R}} \text{ is an } (\tilde{f}^t) \text{ orbit contained in } U\}$ is the set $\widetilde{K} = hK$ associated with $(\tilde{f}^t)$ in Section 15.2.*

Remarks on prehyperbolic sets, similar to those of Section 15.6, could

---

[46] See Hirsch, Palis, Pugh, and Shub [1] The proof in that paper uses normal hyperbolicity to orbits rather than shadowing.

[47] See Pugh and Shub [2]. The proof in that paper again uses normal hyperbolicity to orbits rather than shadowing.

be repeated here. Of course, if (f^t) is an injective semiflow (e.g., a flow), a prehyperbolic set is hyperbolic.

If K is hyperbolic or has local product structure for a flow, the same remains true after time reversal $(t \mapsto -t)$.

16. Homoclinic and Heteroclinic Intersections

In this section, we discuss the intersections between the (global) stable manifold W^- of a hyperbolic fixed point or periodic orbit P and the (global) unstable manifold W^+ of a hyperbolic fixed point or periodic orbit Q. Let $u \in W^- \cap W^+ \backslash (P \cup Q)$. If $P = Q$, we call u a *homoclinic* point; if $P \neq Q$, we call u a *heteroclinic* point. In particular, we shall be interested in the bifurcations that arise when new homoclinic or heteroclinic points are created. Since we shall deal with phenomena that are no longer strictly local, the *global* stable and unstable manifolds $W^\pm$ will be needed instead of the *local* stable and unstable manifolds $V^\pm$ used up to now. We shall thus assume that (f^t) is defined for $t \in \mathbf{Z}$ (dynamical system generated by a diffeomorphism f) or for $t \in \mathbf{R}$ (flow). We also assume that $(x, t) \mapsto f^t x$ is $C^{\mathbf{r}}$ (i.e., we have a $C^{\mathbf{r}}$ diffeomorphism or a $C^{\mathbf{r}}$ flow) with $\mathbf{r} \geqslant 1$. We write

$$W^- = W_P^- = \{x \in M : f^t x \to P \text{ when } t \to +\infty\}$$
$$= \bigcup_{t \geqslant 0} f^{-t} V^-,$$

$$W^+ = W_Q^+ = \{x \in M : f^{-t} x \to Q \text{ when } t \to +\infty\}$$
$$= \bigcup_{t \geqslant 0} f^t V^+,$$

where each $f^{-t} V^-$ or $f^t V^+$ is $C^{\mathbf{r}}$. (See Problem 2 of Part 1 and note in particular that, while $W^\pm$ is locally a manifold, it may fold on itself so as to be dense in $M!$)[48] Thus $u \in W^- \cap W^+$ means that $f^t u \to P$ for $t \to +\infty$ and $f^t u \to Q$ for $t \to -\infty$.

The orbit of a homoclinic point is called a *homoclinic orbit* and obviously consists of homoclinic points; similarly, for a *heteroclinic orbit*. As

---

[48] It would be possible to use a global unstable manifold W^+, but only a local stable manifold V^-. Correspondingly, (f^t) could be defined for $t \geqslant 0$ only, provided, for instance, that $Tf : TM \to TM$ is injective and V^+ finite dimensional (see Problem 2 of Part 1).

we shall see, the existence of a homoclinic orbit implies a complicated orbit structure.

Notice that, with the notation of Sections 8.3, 8.4, we have

$$(16.1) \qquad P \succ u \succ Q \qquad \text{if} \quad u \in W_P^- \cap W_Q^+,$$

and

$$(16.2) \qquad u \sim P \qquad \text{if} \quad u \in W_P^- \cap W_P^+.$$

16.1. Transversal Homoclinic Points

Let $u \in W_P^- \cap W_P^+$, and $T_u W^- + T_u W^+ = T_u M$ (transversality of W^- and W^+ at u). Furthermore, suppose that $T_u W^- \cap T_u W^+ = \{0_u\}$ (origin of $T_u M$) in the case of a diffeomorphism, or that $T_u W^- \cap T_u W^+$ is one-dimensional along the direction of $\frac{d}{dt} f^t u|_{t=0}$ in the case of a flow with periodic orbit P.[49] We then say that u is a *transversal homoclinic point*. Since W^-, W^+ depend continuously on (f^t), a slightly perturbed dynamical system will again have a transversal homoclinic point near u.

Consider for definiteness the case of a diffeomorphism f, and a fixed point P. Since h is homoclinic, we have $f^k u \to P$ when $k \to \pm\infty$. Up to a diffeomorphism, we may thus assume that u is close to P. Notice that W^+ is folded so that successive layers pile up near P (see Fig. 31) and similarly for W^-.[50] To get some understanding of the situation, we follow Smale's idea to put a *horseshoe* in the picture. Choose a chart that identifies a neighborhood of P with $V^- \times V^+$, where $V^\pm$ are local stable and unstable manifolds of P. Let $B_P^-(r)$ and $B_P^+(\delta)$ be balls in $V^\pm$ with $\delta \ll r$, so that $\mathcal{N} = B_P^-(r) \times B_P^+(\varepsilon)$ is very thin along V^-, and $\mathcal{N} \ni u$. For large n, $f^n \mathcal{N}$ will be very thin along V^+. In fact, we can choose n large such that $f^n \mathcal{N} \ni u$. Fig. 31 represents this situation; the intersection $\mathcal{N} \cap f^n \mathcal{N}$ has one connected component

---

[49]These latter conditions follow from the transversality of W^- and W^+ if M is finite dimensional, because $\dim W^- + \dim W^+$ is $\dim_u M$ for a diffeomorphism and $\dim_u M + 1$ when $W^\pm$ are the stable and unstable manifolds of a periodic orbit P for a flow. If P is a fixed point for a flow, $\dim W^- + \dim W^+ = \dim M$, and since $T_u W^- \cap T_u W^+ \ni \frac{d}{dt} f^t u$, the intersection of W^- and W^+ cannot be transversal. (See Section 16.9).

[50]The piling up is described by the inclination lemma (Problem 5 of Part 1).

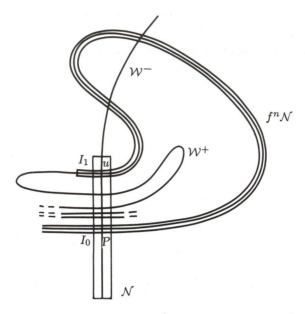

FIG. 31. Homoclinic point $u \in W_P^- \cap W_P^+$. Note the folding of W^+. The image of $\mathcal{N}$ by f^n intersects $\mathcal{N}$ in $I_0 \ni P$ and $I_1 \ni u$.

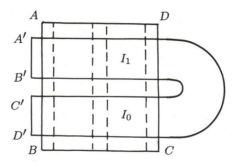

FIG. 32. Smale's horseshoe. In this 2-dimensional abstraction of Fig. 31, $\mathcal{N} = ABCD$, and $g\mathcal{N} = A'B'C'D'$, with $g = f^n$. The boundaries of $g^{-1}\mathcal{N} \cap \mathcal{N}$ are marked by dashed lines.

I_0 around P, and another I_1 around u (other connected components that might be present are disregarded). From now on, we shall write $f^n = g$. The essential features of the situation are abstracted in Fig. 32, which represents *Smale's horseshoe*. In this example, it is assumed for simplicity that M is two-dimensional; we also take g affine in I_0 and I_1 stretching horizontal segments and contracting vertical ones. (These assumptions simplify the analysis but are not necessary.)

It is not hard to see that the set $K = \bigcap_{k \in \mathbf{Z}} g^k \mathcal{N}$ is compact and is a *Cantor set*.[51] In fact, the points x of K are in one-to-one correspondence with the infinite sequences $(\xi_k)_{k \in \mathbf{Z}}$ of 0's and 1's. In this correspondence, ξ_k is 0 or 1 depending on whether $g^k x \in I_0$ or I_1. Two points are close if the corresponding sequences (ξ_k), (η_k) agree for $|k| \leqslant n$, n large. Clearly, the replacement of x by gx corresponds to the *shift* of (ξ_k) by one place to the left. One calls *symbolic dynamics* the dynamics of sequences (ξ_k) defined by the shift, and which replaces the dynamics on K defined by g. The use of symbolic dynamics clarifies many questions. For instance, it is easy to see that periodic points are dense in K. [Equivalently: to every sequence (ξ_k) and $n > 0$, there is a periodic sequence (η_k) such that $\xi_k = \eta_k$ for $|k| \leqslant n$]. Note that the sequence $(\dots, 0, 0, 0, \dots)$ with only zeros and the sequence $(\dots, 0, 1, 0, \dots)$ with a single 1 correspond to P and u, respectively. [Check first that these sequences correspond to points in $\mathcal{W}^- \cap \mathcal{W}^+$.] In particular, there are periodic points for g (hence for f) arbitrarily close to P and u (and different from P). Also, the homoclinic points are dense in K. [Given a sequence (ξ_k) and $n > 0$, there is a sequence (η_k) such that $\eta_k = \xi_k$ for $|k| \leqslant n$ and $\eta_k = 0$ for $|k| > n$.]

The compact set K is a *hyperbolic invariant set* for the diffeomorphism g and has *local product structure*. As indicated in Section 15, hyperbolicity means that there is a continuous splitting $T_K M = V^- + V^+$ of the tangent bundle restricted to K, that $T_g V^\pm = V^\pm$ (invariance), and that there are $C > 0$, $\theta > 1$ with

$$\max_{x \in K} \| T g^{\mp n} | V^\pm \| \leqslant C \theta^{-n} \qquad \text{for } n \geqslant 0.$$

In the horseshoe case of Fig. 32, V^+ is horizontal and V^- vertical, and hyperbolicity is readily verified. In the general case, we use a chart of M at the hyperbolic fixed point P and assume that K is sufficiently close to P. Then the expanding and contracting directions $E^\pm$ at P give almost a hyperbolic splitting of $T_K M$. A true hyperbolic splitting is obtained by using the hyperbolicity criterion 15.1. The local product structure follows from the definition $K = \bigcap_{k \in \mathbf{Z}} g^k \mathcal{N}$ (see Section 15.4).

16.2. Theorem (Smale's homoclinic theorem).[52] *Let f be a C^1 diffeomor-*

---

[51] See Appendix A.3.

[52] The proof follows the outline given above for the case of a fixed point P. We refer

phism and u a transversal homoclinic point for the hyperbolic periodic orbit P. There is then an integer n such that $g = f^n$ has a hyperbolic compact invariant set K homeomorphic to a Cantor set and containing P and u. The periodic points and the homoclinic points (for P) are dense in K.

If P is a fixed point, $g|K$ is topologically conjugate to the shift acting on the space $\{I_0, I_1\}^{\mathbf{Z}}$ of binary sequences. If P has period N, there is an N to 1 map $\pi : K \mapsto \{I_0, I_1\}^{\mathbf{Z}}$ such that $\pi \circ (g|K) = \text{shift} \circ \pi$.

Notice that there is some arbitrariness in the choice of K in the theorem. Since the homoclinic points are dense in K, (16.2) shows that $K \subset [P]$, where $[P]$ is the basic class of P (see Section 8.4). The invariant set $[P]$ may be hyperbolic or not and may be an attractor or not. On the other hand, the set K is hyperbolic but cannot be an attractor: One can choose $x \in W_P^+ \backslash K$, and therefore $K \succ x$, but $x \notin K$. (Similarly, Smale's horseshoe is not an attractor.)

16.3. The Case of Flows

Let (f^t) be a C^1 flow, i.e., $(x,t) \mapsto f^t x$ is C^1, and let u be a transversal homoclinic point for the hyperbolic periodic orbit P. The flow then has a hyperbolic compact invariant set[53] K containing P and u. The periodic orbits and the homoclinic orbits for P are dense in K. There is a C^1 local cross-section Σ of the flow such that $K_0 = K \cap \Sigma$ is a Cantor set, and a continuous function $\tau > 0$ on K_0 such that $K = \{f^t x : x \in K_0, 0 \leqslant t < \tau(x)\}$.

To see this, one notices that the problem is essentially a local one and can be studied by a Poincaré map near a point of P. (The orbit of u wanders away from P only for a limited amount of time.)

16.4. Corollary. *If u is a transversal homoclinic point for a periodic orbit of a diffeomorphism or flow, u is in the closure of the hyperbolic periodic orbits.*

16.5. Homoclinic Tangencies

Suppose that a homoclinic intersection is created when we increase a bifurcation parameter $\mu \in \mathbf{R}$ past the value μ_0. This bifurcation has

to Smale [2], Moser [1], and Newhouse [3] for details.

[53] See Section 15.

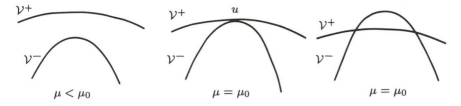

FIG. 33. Creation of a nondegenerate tangency. If a is a periodic point for the 2-dimensional diffeomorphism f_μ, a codimension 1 bifurcation occurs when $\mathcal{V}^- = \mathcal{V}_a^-(\mu)$ intersects $\mathcal{V}^+ = \mathcal{V}_{f^k a}^+(\mu)$.

been much studied in the case of a two-dimensional diffeomorphism, and we shall restrict our discussion to this case (although most results can be extended to higher dimension).

Let thus $\dim M = 2$, and $f_\mu : M \mapsto M$ be a C^2 diffeomorphism such that $(\mu, x) \mapsto f_\mu x$ is C^1 (we shall make stronger assumptions later). We assume that $P = P(\mu)$ is a saddle-type periodic orbit of period N for $f = f_\mu$, so that $\dim W^-(\mu) = \dim W^+(\mu) = 1$ and $P = \{a, fa, \ldots, f^{N-1}a\}$. We let $\mathcal{V}_a^-(\mu), \mathcal{V}_{f^k a}^+(\mu)$ be (suitably large) local stable and unstable manifolds. The homoclinic point $u \in \mathcal{V}_a^-(\mu_0) \cap \mathcal{V}_{f^k a}^+(\mu_0)$ created when μ reaches μ_0 cannot be transversal (because transversal intersections persist under perturbations). Generically, however, $\mathcal{V}_a^-(\mu_0)$ and $\mathcal{V}_{f^k a}^+(\mu_0)$ have a *nondegenerate tangency* at u. This means that if $\mathcal{V}^\pm$ are the graphs of maps $\xi \mapsto \eta = \varphi_\mu^\pm(\xi)$ in a chart centered at u, and we write $\varphi_\mu = \varphi_\mu^+ - \varphi_\mu^-$, then φ_{μ_0} has a nonvanishing second derivative (see Fig. 33). Generically also, the tangency is *crossed at nonzero speed*, i.e., $\frac{\partial}{\partial \mu}\varphi_\mu(0)|_{\mu=\mu_0} \neq 0$. [Our differentiability assumptions on f_μ imply that $\xi \mapsto \varphi_\mu(\xi)$ is C^2, and $(\mu, \xi) \mapsto \varphi_\mu(\xi)$ is C^1. The latter fact is seen by noting that the stable and unstable manifolds of the normally hyperbolic manifold $x = a(\mu)$ under the C^1 diffeomorphism $(\mu, x) \mapsto (\mu, f_\mu(x))$ are C^1.]

Since we assume that a homoclinic intersection is created when μ *increases* past μ_0, Fig. 33 corresponds to $\varphi_{\mu_0}(0) = 0$, $\varphi'_{\mu_0}(0) = 0$, $\varphi''_{\mu_0}(0) > 0$, and $\frac{\partial}{\partial \mu}\varphi_\mu(0)|_{\mu=\mu_0} < 0$. There are thus an $\varepsilon > 0$ and a neighborhood U of u such that

(a) $\mathcal{V}_a^-(\mu) \cap \mathcal{V}_{f^k a}^+(\mu) \cap U = \emptyset$ for $\mu \in (\mu_0 - \varepsilon, \mu_0)$.

(b) $\mathcal{V}_a^-(\mu)$ and $\mathcal{V}_{f^k a}^+(\mu)$ have two transversal intersection points in U for $\mu \in (\mu_0, \mu_0 + \varepsilon)$.

16.6. Theorem (Creation of Infinitely Many Sinks in a Homoclinic Tangency).[54] *Let* $\dim M = 2$, *and* (f_μ) *be a one parameter family of* C^3 *diffeomorphisms of* M, *with* C^1 *dependence*[55] *on* μ. *Suppose that a is a hyperbolic periodic point*[56] *of period N for f_μ, such that* $|\det(T_a f_{\mu_0}^N)| < 1$. *Suppose also that* $V_a^-(\mu)$, $V_a^+(\mu)$ *are 1-dimensional, have a nondegenerate homoclinic tangency at u for $\mu = \mu_0$, and that this tangency is crossed at nonzero speed*[57] *when μ increases past μ_0.*

Under these conditions, for each choice of $\varepsilon > 0$, there is a nonempty interval $(\mu_1, \mu_2) \subset (\mu_0, \mu_0 + \varepsilon)$ and a residual subset $J \subset (\mu_1, \mu_2)$ such that, for $\mu \in J$, f_μ has infinitely many attracting periodic orbits (sinks).

Residual sets and genericity have been introduced in Section 8.7. The above theorem is a difficult one, and we shall only discuss the ideas on which the proof is based.

16.7 Wild Hyperbolic Sets (Newhouse theory)[58]

Let f be a C^r diffeomorphism of the 2-dimensional manifold M, $r \geq 2$, and let K be an invariant set with local product structure for f (see Section 15.4). We assume that the stable and unstable dimensions for K are both equal to 1, and that f restricted to K is *topologically transitive*.[59] In view of the persistence property of sets with local product structure (Section 15.6), a hyperbolic set $K(g)$ close to K is defined for diffeomorphisms g close to f. Suppose that there is a neighborhood $\mathcal{N}$ of f for the C^r topology such that when $g \in \mathcal{N}$ there are $x, y \in K(g)$ for which the global unstable manifold W_x^+ and stable manifold W_y^- are tangent somewhere. Then, K is called a *wild hyperbolic* set for f.

(Existence of wild hyperbolic sets). *Let a be a periodic point of period N for the C^r diffeomorphism $f : M \mapsto M$, with $r \geq 2$ and M compact*

---

[54]This result is essentially due to Newhouse; an explicit proof of the above statement is in Robinson [3].

[55]We assume thus that $D_\mu^k D_x^\ell f_\mu(x)$ is continuous for $k = 0, 1$; $\ell = 0, 1, 2, 3$.

[56]The point a depends smoothly on μ and could be fixed by a change of coordinates.

[57]This condition could be weakened to conditions (a) and (b) of Section 16.5.

[58]See Newhouse [1], [2], [3].

[59]This means that there is $x \in K$ such that $\{f^n x : n \in \mathbf{Z}\}$ is dense in K (see Appendix C.1). Topologically transitive sets with local product structure are sometimes called *hyperbolic basic sets*. We do not use this terminology, which conflicts with our notion of basic classes.

2-dimensional. *Assume that a is hyperbolic and that the corresponding stable and unstable manifolds are of dimension 1 and tangent at some point u; also, let $|\det(T_a f^N)| < 1$. Then arbitrarily C^r close to f, there is a C^r diffeomorphism g having a wild hyperbolic set near the orbit of u.*

(Infinitely many sinks). *Let $f : M \mapsto M$ be a C^r diffeomorphism, with $2 \leqslant r < \infty$ and M compact 2-dimensional, such that f has a wild hyperbolic set K containing a point a of period N with $|\det(T_a f^N)| < 1$. There are then an open neighborhood $\mathcal{N}$ of f in the C^r diffeomorphisms of M and a residual subset $\mathcal{R}$ of $\mathcal{N}$ such that, if $g \in \mathcal{R}$, the closure of the set of attracting periodic points contains $K(g)$.*

16.8. Outline of the Proof of Theorem 16.6

The idea if to adapt the results just stated to one-parameter families of diffeomorphisms.

(A) *Under the assumptions of Theorem 16.6, for each $\varepsilon > 0$ there is a nonempty interval $(\mu_1, \mu_2) \subset (\mu_0, \mu_0 + \varepsilon)$ such that, for $\mu \in (\mu_1, \mu_2)$, F_μ has a wild hyperbolic set $K_\mu \ni a$.*

For $\mu > \mu_0$, there will be a transversal homoclinic intersection, and therefore (by Smale's Homoclinic Theorem 16.2), a hyperbolic Cantor set K_μ containing a. The problem is to choose K_μ such that there is a persistent tangency between the stable and unstable manifolds of K_μ (see Fig. 34). To handle this, Newhouse introduced the key concept of *thickness* τ of a linear Cantor set (τ is somewhat related to the *Hausdorff dimension*[60]). Given Cantor sets $A, B \subset \mathbf{R}$, Newhouse gives a criterion based on thickness that ensures that $A \cap B \neq \emptyset$ (see Problem 6 for details). The criterion is applied to the bunch of stable and the bunch of unstable manifolds of K_μ (which are Cantor sets in cross-section) to obtain persistent tangencies. For many μ close to μ_0, one can construct K_μ such that the stable and unstable bunches are suitably *thick*. The construction of K_μ uses the closeness to a tangency of $\mathcal{V}_a^-$ and $\mathcal{V}_a^+$ and requires a delicate geometrical and analytical discussion for which we must refer the reader to the original papers.

(B) *Suppose that, for each $\mu \in (\mu_1, \mu_2)$, f_μ has a wild hyperbolic set containing a with $|\det(T_a f_{\mu_0}^N)| < 1$. There is then a residual set $J \subset (\mu_1, \mu_2)$ such that for $\mu \in J$, f_μ has infinitely many sinks.*

---

[60] See Appendix A.3.

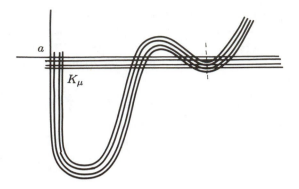

FIG. 34.   Persistence of tangencies. If the bunch of stable and the bunch of unstable manifolds of the hyperbolic set K_μ are sufficiently thick, there are persistent tangencies of the stable and unstable manifolds.

For simplicity, take a to be a fixed point. The creation of one sink goes as follows. One proves geometrically that if W_a^- and W_a^+ have a tangency at μ_0 such that locally they don't cross for $\mu < \mu_0$ and do cross for $\mu > \mu_0$, then f_μ has, for some $\mu > \mu_0$ and arbitrarily close to μ_0, a periodic point c of period n such that $T_c f_\mu^n$ has an eigenvalue $\lambda_1 < -1$. For lower values of μ, f_μ^n has no fixed point in the region of c, and therefore c could only be created by a saddle-node bifurcation.[61] From using $|\det T_c f_\mu^n| < 1$, it follows that for some intermediate values of μ (forming an open interval) c is a sink. In the situation of Fig. 34, this process occurs often because the global stable (resp. unstable) manifold of a is dense in the bunch of stable (resp. unstable) manifolds of K_μ. In this manner, one obtains infinitely many sinks for a residual set of values of μ in (μ_1, μ_2).

16.9. A Theorem of Palis and Takens

The study of homoclinic tangencies opens the way to a *geometric* understanding of *nonhyperbolic* dynamical systems. This explains that a lot of hard work has been devoted to understanding such tangencies. Here we shall describe a theorem of Palis and Takens,[62] which in a sense is

---

[61] Not necessarily a generic saddle-node bifurcation.

[62] See Palis and Takens [3], and for related earlier work, Newhouse and Palis [2], Palis and Takens [2]. I am indebted to J. Palis for a description of the results in the form presented here.

complementary to the results of Newhouse described earlier.

Let M be a compact 2-dimensional manifold and (f_μ) a sufficiently smooth one-parameter family of diffeomorphisms of M defined for μ in an interval around 0. We assume that f_0 has a compact invariant set K_0 with local product structure, such that the stable and unstable dimensions for K_0 are both equal to 1, and that f_0 restricted to K_0 is topologically transitive. For μ near 0, there is then a set K_μ with the same properties and depending continuously on μ. Suppose that the bunches of stable and unstable manifolds of K_μ intersect only at K_μ for $\mu < 0$, but that a tangency is created at $\mu = 0$. One can show that such a tangency necessarily occurs between the stable and unstable manifolds of a *periodic orbit*.

Let thus $a_\mu \in K_\mu$ be a periodic point for f_μ, depending continuously on μ, and such that there is a unique orbit $\{f_0^n u\}$ of homoclinic tangency, associated with the orbit of $a = a_0$. With the notation and terminology of 16.5, we assume that $\mathcal{V}_a^-(0)$ and $\mathcal{V}_{f^k a}^+(0)$ have a non-degenerate tangency at u and that this tangency is crossed at nonzero speed when μ increases past 0.

The crucial assumption of Palis and Takens is that the stable and unstable bunches emerging from K_0 are sufficiently *thin*. Specifically, it is required that

(16.3) $\dim_H(K_0 \cap \mathcal{V}_a^+(0)) + \dim_H(K_0 \cap \mathcal{V}_a^-(0)) < 1,$

where $\dim_H$ denotes Hausdorff dimension (see Appendix A.3). *If* (16.3) *holds, one can find an open set* $V \supset K_0 \cup (\bigcup_k f^k u)$ *such that if* Λ_μ *is the union of the nonwandering orbits for* f_μ *contained in* V, *then*

$$\lim_{\delta \to 0} \frac{m\{\mu \in [0, \delta] : \Lambda_\mu \text{ is hyperbolic for } f_\mu\}}{\delta} = 1,$$

where m denotes Lebesgue measure.

What this says is that, under condition (16.3), the set of nonwandering points locally created after the homoclinic tangency is hyperbolic for most values of μ. This result is complementary to Theorem 16.6, which says that there is a set of values of μ that is large in some sense, and for which the nonwandering set is not hyperbolic.

16.10. Saddle Connections

Let (f^t) be a flow on a two-dimensional manifold M with fixed points a, b of saddle type (we allow a and b to coincide). A *saddle connection*

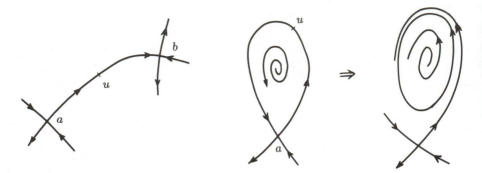

FIG. 35.   A saddle connection is an orbit connecting two saddle-type fixed points a and b of a 2-dimensional flow. If $a = b$, we have a homoclinic orbit that may bifurcate to an attracting or repelling periodic orbit of long period.

is an orbit $\Gamma = \{f^t u\}$ such that $\lim_{t \to -\infty} f^t u = a$, $\lim_{t \to +\infty} f^t u = b$. This means that a piece of unstable manifold of a coincides with a piece of stable manifold of b (see Fig. 35). This situation is nongeneric and corresponds to a bifurcation.

In particular, if $a = b$, Γ is a homoclinic orbit that may create an attracting or repelling periodic orbit by a codimension 1 bifurcation (Γ is a "periodic orbit of infinite period" and creates a periodic orbit of long period).

16.11. Flow with Fixed Point Having a Homoclinic Orbit (Šil'nikov Theory)

Let (f^t) be a flow on a manifold M of any finite dimension, and a a fixed point of (f^t). A homoclinic point u corresponding to a cannot be transversal (see Section 16.1). Therefore, the existence of the homoclinic orbit $\{f^t u\}$ corresponds to a bifurcation.

Let (f^t) be obtained by integrating a vector field X. Following Šil'nikov,[63] we consider the case where $\dim M = 3$, and $T_a X$ has eigenvalues χ_1, χ_2, and χ_2^*, where

(16.4) $\chi_1 > -\operatorname{Re} \chi_2 > 0$, $\operatorname{Im} \chi_2 > 0$.

If X is of class $C^{(1,1)}$ we claim that *there are infinitely many periodic orbits of saddle type close to the homoclinic orbit* $\Gamma = \{f^t u\}$.

[63]See Šil'nikov [1], Tresser [1], [2], and references quoted there.

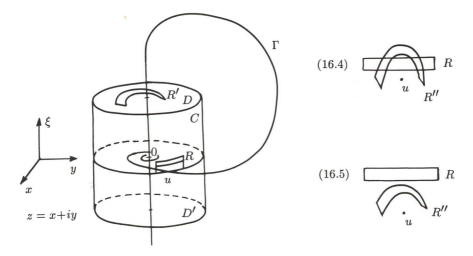

FIG. 36. Homoclinic orbit Γ of a fixed point 0 for a flow in $\mathbf{R}^3$. The occurence of this homoclinic orbit is nongeneric. By using a Poincaré map, one obtains in case (16.4) an infinity of periodic orbits near Γ, in case (16.5) no periodic orbit near Γ.

In view of Section 5.8, we may linearize the flow in a neighborhood of a. Taking a chart that identifies a with the origin 0 of $\mathbf{R} \times \mathbf{C}$, we have thus

$$f^t \begin{pmatrix} \xi \\ z \end{pmatrix} = \begin{pmatrix} e^{\chi_1 t} & \xi \\ e^{\chi_2 t} & z \end{pmatrix}$$

in a neighborhood of 0. We enclose 0 by a piece of circular cylinder C and two disks D, D'; we may assume that $u \in C$ (see Fig. 36). Following the orbits of (f^t) inside the cylinder, we have a map of the upper part of C to D and of the lower part to D'. This map sends horizontal circles of C to circles of D or D', and vertical lines of C to logarithmic spirals. A Poincaré map can be defined from part of C to C. It sends a rectangle R close to the point of intersection u of C with Γ first to a figure R' bounded by two arcs of circles and two arcs of logarithmic spirals in D, then to $R'' \subset C$. Since R'' is a slightly distorted version of R', it can be assumed to turn around u many times, and to intersect R because of the assumption (16.4). By making R thinner, one obtains a horseshoe, and therefore, (f^t) has a hyperbolic invariant set containing an infinity of saddle-type periodic orbits (see Section 16.1).

Although the homoclinic orbit Γ may be destroyed by a small per-turbation, notice that the invariant set is persistent (see Sections 15.5, 15.6).

If the condition (16.4) is replaced by

(16.5) $-\operatorname{Re}\chi_2 > \chi_1 > 0, \qquad \operatorname{Im}\chi_2 > 0,$

the horseshoe R'' misses the rectangle R if R is sufficiently thin and close to U. Therefore, there are no periodic orbits near Γ.[64]

17. Global Bifurcations

There is no such thing as a general theory of bifurcations. However, one can hope to understand a number of typical phenomena, and apply this knowledge to the discussion of dynamical systems that occur in physical experiments or computer simulations. We have thus far approached the bifurcation problem from a mainly *local* point of view. In this section, we shall make some general and qualitative remarks on *global* bifurcations.

We use the language of the relation $\succ$ of Sections 8.3, 8.4. This yields a somewhat coarse description of the dynamical system because a given basic class $[a]$ contains points related by pseudoorbits rather than by true orbits of the system. Such a coarse description is appropriate because of the enormous complication that dynamical systems may exhibit.

17.1. Changes in Order Structure: Cycles

Consider a dynamical system depending continuously on a real parameter μ. When $\mu_n \to \mu_0$, let $a_n \to a$, $b_n \to b$, then $a_n \succ b_n$ implies $a \succ b$. In particular, $\lim[a_n] \subset [a]$. If we have an order structure independent of n on the classes $[a_n]$, we have an order-preserving map

$$[a_n] \mapsto [a].$$

The bifurcation at μ_0 may thus collapse several basic classes into one and also create new basic classes. If a new basic class is created, it cannot be an attractor; also, an attractor need not be mapped to an attractor (see Fig. 37).

The saddle-node bifurcation gives an example of creation of a basic class or an example of collapse (depending on whether the μ_n increase

---

[64]Under small perturbations of the flow, an attracting periodic orbit may or may not appear (see Šil'nikov [1], Tresser [2]).

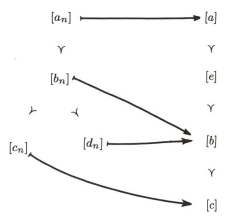

FIG. 37. Change of order structure of basic classes in a bifurcation. The classes $[b_n], [d_n]$ collapse into $[b]$, and a new basic class $[e]$ is created.

or decrease to μ_0). The flip and the Hopf bifurcations give examples of collapse.

Suppose that $[a_n] \succ [b_n] \succ \cdots \succ [e_n]$ and $a_n \to a, b_n \to b, \ldots, e_n \to e$. If $e \succ a$, we have created a *cycle*, and it is clear that $a \sim b \sim \cdots \sim e$. Furthermore, the points x between two members of the chain (for instance $a \succ x \succ b$) are now also in $[a]$. This corresponds to an *explosion* of the chain-recurrent set. For instance, if $[a_n]$ and $[e_n]$ are hyperbolic periodic orbits, the manifolds $\mathcal{W}_{e_n}^+$ and $\mathcal{W}_{a_n}^-$ are disjoint (see (16.1)), and if a heteroclinic point $u \in \mathcal{W}_e^+ \cap \mathcal{W}_a^-$ appears, then an explosion of the chain-recurrent set occurs, and $u \in [a]$. It may happen that the nonwandering set at the value μ_0 of the bifurcation parameter is the union of hyperbolic sets $\Omega_a = \lim[a_n], \ldots, \Omega_e = \lim[e_n]$, forming a cycle. In that case, the nonwandering set Ω is strictly smaller than the chain-recurrent set, and small perturbations may cause an explosion of the nonwandering set or Ω-*explosion*.[65]

17.2. Use of Larger Invariant Sets

The order structure of basic classes is often unstable under perturbations and varies rapidly with μ. It is then useful to consider bigger closed invariant sets for which the order $\succ$ is still defined. (These bigger sets contain basic classes and possibly points that are not chain recurrent).

---

[65] See Palis [3], [4].

Examples are the attracting sets of Section 8.1, and the normally hyperbolic manifolds of Section 14.[66] Numerical studies of strange attractors, in fact, usually involve attracting sets of rapidly changing nature.

17.3. Changes in the Basins of Attraction

Let U be a nonempty open set such that the closure of $\bigcup_{\mu \in J} f_\mu^t U$ is compact and contained in U for some interval J of values of μ and all sufficiently large t. We have seen (Proposition 8.2) that

$$\Lambda_\mu = \bigcap_{t \geqslant 0} f_\mu^t U$$

is a compact attracting set with fundamental neighborhood U. The basin of attraction of Λ_μ is

$$W_\mu = \bigcup_{t \geqslant 0} (f_\mu^t)^{-1} U.$$

Every limit point of Λ_μ when $\mu \to \mu_0$ is a point of Λ_{μ_0} (upper semicontinuity of Λ_μ). Every point of W_{μ_0} belongs to W_μ for μ close to μ_0 (lower semicontinuity of W_μ). These semicontinuity properties imply some kind of persistence of Λ_μ (remember also that $\Lambda_\mu \neq \emptyset$). The destruction of the attracting set Λ_μ may occur when Λ_μ touches the boundary of its basin, after which this basin may empty itself into another basin of attraction. This is a *catastrophe* in the sense of Thom [1].[67] (Catastrophe theory is devoted to the study of such bifurcation for *gradient flows*, defined in Appendix B.7.)

17.4. The Boundary of a Basin of Attraction

Let us assume that the manifold M is compact. *For each $x \in M$, there is a unique basic class $[a] = \omega(x)$ such that $d(f^t x, [a]) \to o$ when $t \to +\infty$.* [If a, b are any two limits of $f^t x$ when $t \to +\infty$, then $a \succ b \succ a$.][68] We have thus a partition of M into sets $\omega^{-1}[a]$. The basin of an attracting set Λ is the union of the $\omega^{-1}[a]$ with $[a]$ contained in Λ. The boundary of a basin is also a union of sets $\omega^{-1}[a]$. In particular, pieces of

---

[66] The persistence of attracting sets is discussed in Section 17.3; the persistence of normally hyperbolic manifolds was discussed in Section 14.2.

[67] Grebogi, Ott, and Yorke [1] speak of *boundary crises*; see their paper for the discussion of examples involving strange attractors.

[68] This is also true in the situation of Proposition 8.2, with x in the basin of attraction of Λ.

the boundary may be constituted by stable manifolds of periodic orbits (or of hyperbolic sets or normally hyperbolic manifolds). For instance, the stable manifold of a fixed point of saddle type in 2 dimensions may separate the basins of two attracting sets (*separatrix*). Notice that such a stable manifold tends to fold, with the folds accumulating on an unstable manifold (see Fig. 31). Therefore, the boundary of a basin of attraction is typically very convoluted and complicated; it can also be nondifferentiable.[69] The basic classes contained in the boundary of a basin of attraction may undergo bifurcations, which are of limited physical interest if they do not affect the attractors. The convolution of the boundaries of basins of attraction is physically significant, however, as already appreciated by Poincaré [4]. It means that a small change in initial condition may send an orbit towards one attractor rather than the other.[70] This new kind of *sensitivity to initial condition* (quite different from that discussed in Section 8.4) is frequently seen in computer studies.[71]

17.5. Changes in the Internal Structure of Attracting Sets

We have seen that a homoclinic tangency may create infinitely many sinks for a residual set of values in an interval of the bifurcation parameter (Theorem 16.6). This gives an example of an attracting set that is very unstable under perturbations and changes rapidly with μ. It is of course not possible to follow the infinitely many changes that the attracting set undergoes as μ varies. It is nevertheless useful to interpret certain gross changes visible in numerical simulations. Such a change occurs, for instance, when a strange attractor hits a saddle-type periodic orbit in its domain of attraction and then suddenly explodes into a bigger attractor.[72]

It should become increasingly clear to the reader that, in general, it is

---

[69] The complex map $z \mapsto z^2 + \alpha z$ has the attracting fixed points 0 and ∞ on the Riemann sphere, and their basins are separated by a curve with Hausdorff dimension > 1 if α is small and $\neq 0$ (see, for instance, Ruelle [6]).

[70] If a magnetic pendulum (with some friction) is suspended in the field created by a few magnets, it undergoes complicated oscillations before coming to rest in an equilibrium position. The domains of attraction of the various equilibria have very complicated boundaries here, and it is hard to guess from the initial data what will be the final position.

[71] See, for instance, Grebogi, McDonald, Ott, and Yorke [1].

[72] By definition, the saddle-type periodic orbit is part of the attracting set, but it would not be visible as part of the strange attractor in a numerical study. The sudden change in the attractor is an *interior crisis* in the sense of Grebogi, Ott, and Yorke [1].

not possible to understand the behavior of a dynamical system, or its bi-furcations, by just staring at the defining equations. On the other hand, an interpretation of computer studies in terms of known bifurcations is often possible and useful. If desired, it may be possible to transform some computer observations into proofs. (For instance, if a homoclinic tangency is observed, a computer-assisted proof of this fact should also be possible.)

Note

Part 2 of this monograph discusses bifurcation theory, focussing attention first on fixed points[73] and periodic orbits.

Sections 9–13 deal with problems that can be studied locally near a fixed point or periodic orbit, namely the saddle-node, flip, and Hopf bifurcations. We study maps and semiflows in parallel. By using a center manifold, one reduces to the case of a dynamical system in 1 or 2 dimensions.

Two general tools for the study of differentiable dynamical systems are inserted in Sections 14 and 15. These are the normally hyperbolic invariant manifolds (which are needed in the study of the Hopf bifurcation) and the hyperbolic invariant sets. Both concepts extend hyperbolicity from fixed points or periodic orbits to more complicated sets. The standard reference for these questions is Hirsch, Pugh, and Shub [1].

Hyperbolic sets have ancient origins. Geodesic flows on surfaces of constant negative curvature, studied by Hadamard, constitute an important example. Major steps towards the understanding of hyperbolic sets are due to Anosov [1] and Smale [3]. The latter gave a general definition, and introduced local product structure and the *Axiom A*, to be discussed in Appendix D. Bowen developed the systematic use of shadowing. For more detailed historical remarks, see Anosov [1] and the "commentaires" of Shub [2]. In the present monograph, we systematically generalize hyperbolicity to infinite dimension on one hand, to maps and semiflows (rather than diffeomorphisms and flows) on the other hand.

Many interesting phenomena occur when the stable manifold of a fixed point or periodic orbit intersects the unstable manifold of another,

---

[73] In certain circumstances, the existence of fixed points can be obtained from topological index arguments; see Appendix B.2.

or the same, fixed point or periodic orbit. In particular, Section 16 discusses Smale's horseshoe and the homoclinic theorem, Newhouse's wild hyperbolic sets and the associated infinitely many sinks, and Šil'nikov's theory of a flow with homoclinic orbit. The phenomena encountered here give an idea of the extremely complicated global behavior that differentiable dynamical systems may exhibit. A rough description of global bifurcations is given in Section 17.

General references to the contents of this chapter have already been given in the Note to Part 1.

Problems

1. **(Applications of Transversality).**

   (a) Let V and M be compact manifolds. Show that $\mathcal{F} = \{(\mu, x) \in V \times M : f(\mu, x) = x\}$ is a $C^{\mathbf{r}}$ submanifold of $V \times M$ for f in a dense open subset of $C^{\mathbf{r}}(V \times M, M)$, when $\mathbf{r} \geqslant 1$. In particular, by taking V to be 1-dimensional (a circle), justify the genericity considerations of Section 9.1.

   [Let $\mathcal{D} \subset C^r(V \times M, M)$ consist of those f such that the map $(\mu, x) \mapsto (f(\mu, x), x)$ of $V \times M$ to $M \times M$ is transversal to the diagonal $Y = \{(x, x) : x \in M\}$. Clearly, $\mathcal{D}$ is open. For any $f \in C^{\mathbf{r}}(V \times M, M)$, use now Thom's lemma B.4.6 to approximate $(\mu, x) \mapsto (f(\mu, x), x)$ by $(\mu, x) \mapsto (g_\mu(x), h_\mu(x))$ transversal to Y. Check that $(\mu, x) \mapsto (h_\mu^{-1} \circ g_\mu(x), x)$ is again a transversal approximation, so that $\mathcal{D}$ is dense. If $f \in \mathcal{D}$, the smoothness of $\mathcal{F}$ follows from Proposition B.4.4.]

   (b) Similarly, let $\mathcal{X}^{\mathbf{r}}$ be the space of vector fields (X_μ) on M depending on a parameter $\mu \in V$ and such that $(\mu, x) \mapsto X_\mu(x)$ is $C^{\mathbf{r}}$ on $V \times M$. The set $\mathcal{F} = \{(\mu, x) : X_\mu(x) = 0\}$ is a $C^{\mathbf{r}}$ submanifold of $V \times M$ for a dense subset of $\mathcal{X}^{\mathbf{r}}$, with the natural topology and $\mathbf{r} \geqslant 1$.

   [Approximate $(\mu, x) \mapsto X_\mu(x)$ by a map $V \times M \mapsto TM$ that is transversal to the zero-section of TM.]

2. **(Reformulation of Normal Hyperbolicity).** Let H be a compact set, V a (Banach) vector bundle over H with subbundle $W \subset V$, and let $q_x : V_x \mapsto V_x/W_x$ be the quotient map. We denote by $f : H \mapsto H$ a homeomorphism, by $T : V \mapsto V$ a vector bundle map over f such that $TW \subset W$, and by $\tilde{T}$ the bundle map associated with T on the quotient

$\tilde{V} = V/W$ (i.e., $\tilde{T}_x = q_{fx}Tq_x^{-1}$). We assume that there is a continuous splitting $\tilde{V} = \tilde{V}^- + \tilde{V}^+$ such that $\tilde{T}\tilde{V}^- \subset \tilde{V}^-$, $\tilde{T}\tilde{V}^+ = \tilde{V}^+$; we denote by $\tilde{T}^\pm$ the restriction of $\tilde{T}$ to $\tilde{V}^\pm$ and by T^0 the restriction of T to W. We also assume that the $T_x^0 : W_x \mapsto W_{fx}$ are invertible and that there are $R \geqslant 1$, $\tilde{C} \geqslant 1$, and $\theta > 1$ such that

(1)
$$\frac{\|(\tilde{T}^-)_x^n\|}{m((T^0)_x^n)^k} \leqslant \tilde{C}\theta^{-n}, \qquad \frac{m((\tilde{T}^+)_x^n)}{\|(T^0)_x^n\|^k} \geqslant \tilde{C}^{-1}\theta^n$$

whenever $0 \leqslant k \leqslant R$ and $n \geqslant 0$.

(a) If there is a continuous splitting $V = W + X$, show that a continuous splitting $V = W + V^- + V^+$ exists, with $C > 1$, such that $TV^- \subset V^-$, $TV^+ = V^+$, and

$$\frac{\|(T^-)_x^n\|}{m((T^0)_x^n)^k} \leqslant C\theta^{-n}, \qquad \frac{m((T^+)_x^n)}{\|(T^0)_x^n\|^k} \geqslant C^{-1}\theta^n$$

whenever $0 \leqslant k \leqslant R$ and $n \geqslant 0$. Here $T^\pm$ is the restriction of T to $V^\pm$, and naturally $\tilde{V}_x^\pm = q_x V_x^\pm$

[By assumption there is a linear map (projection) $p_x : V_x \mapsto X_x$ such that $p_x W_x = 0$ and p_x is the identity on X_x. Let $X_x^\pm = X_x \cap q_x^{-1}\tilde{V}_x^\pm$, and determine the $V_x^\pm$ as graphs of linear maps $A_x^\pm : X_x^\pm \mapsto W$ satisfying the invariance condition

$$A_{f^n x}^\pm[p_{f^n x}T_x^n(u + A_x^\pm u)] = (1 - p_{f^n x})T_x^n(u + A_x^\pm u),$$

or, since $p_{f^n x}T_x^n A_x^\pm = 0$,

$$A_x^\pm = (T_x^n)^{-1}A_{f^n x}^\pm p_{f^n x}T_x^n - (T_x^n)^{-1}(1 - p_{f^n x})T_x^n.$$

Taking $k = 1$ in (1), choose n such that

$$\max_x \|(T_x^n)^{-1}|W_{fx}^n\| \cdot \|p_{f^n x}T_x^n|X_x^-\| < 1$$

and prove the existence of a continuous function $x \mapsto A_x^-$ by a contraction mapping argument.[74] Since $\tilde{T}^+$ is invertible, one may replace f by f^{-1} to get $x \mapsto A_x^+$ by a similar argument.]

[74]See Fenichel [1], Theorem 6; Hirsch, Pugh, and Shub [1], Proposition (2.3).

(b) Prove Remark 14.3 (b) by applying (a) to the case where H is a smooth submanifold of M, f a diffeomorphism of H extending to a smooth map $M \mapsto M$, the bundles V, W are the tangent bundles $T_H M$, TH, and T is Tf restricted to $T_H M$.
[The existence of a continuous splitting $V = W + X$ is ensured by the fact that W has finite dimension.]

(c) Extend (b) to the case of a semiflow.

3. (Proof of the Hyperbolicity Criterion 15.1). Let f be a homeomorphism of the compact set K, and $T : E \mapsto E$ a continuous vector bundle map over f. In this more general setting (T replacing Tf), we assume that the approximate hyperbolicity condition of Section 15.1 holds, and let $\xi_1, \xi_2 \in S^+$. If $\eta_\alpha = T^{km}\xi_\alpha$, $k \geqslant 1$, and $\eta_2 - \eta_1 \in E^-$, we have

$$\|\eta_2 - \eta_1\| \leqslant \lambda^{-k}\|\xi_2 - \xi_1\| \leqslant \lambda^{-k}[\|\xi_1\| + \|\xi_2\|]$$
$$\leqslant \lambda^{-2k}[\|\eta_1\| + \|\eta_2\|] \leqslant \lambda^{-2k}[2\|\eta_1\| + \|\eta_2 - \eta_1\|]$$

so that $\|\eta_2 - \eta_1\| \leqslant 2\lambda^{-2k}(1 - \lambda^{-2k})^{-1}\|\eta_1\|$. Therefore, when $k \to +\infty$, $T^{km}S^+(f^{-km}x)$ tends to a linear space V_x^+. Clearly, T^m maps V_x^+ one to one onto $V_{f^m x}^+$. The conditions $T^m E^+ \subset S^+$, $T^m E^+ + E^- = E$ imply that $T^m E^+$ is the graph of a continuous linear map $A : E^+ \mapsto E^-$ (at each $x \in K$). Also, T^m (graph sA) is the graph of a map $A_s : E^+ \mapsto E^-$ for $s = 0$, and by continuity for $s \in [0,1]$. In particular, $T^{2m} E^+$ is the graph of a map $E^+ \mapsto E^-$. By repeating the argument, we see that for every integer $k \geqslant 1$, $T^{km} E^+$ is the graph of a continuous linear map $A^{(k)} : E^+ \mapsto E^-$. Therefore, E_x^-, V_x^+ are complementary in E_x.

Similarly, check that when $k \to \infty$, $T^{-km}S^-(f^{km}x)$ tends to a linear space V_x^-. From $E^- + V^+ = E$, we get $T^{-km}E^- + V^+ = E$, and therefore V^- and V^+ are complementary in E. We have $T^m V^+ = V^+$, $T^{-m}V^- = V^-$, and

$$\|T^m \xi\| \geqslant \lambda\|\xi\| \qquad \text{if } \xi \in V^+,$$
$$\|T^m \xi\| \leqslant \lambda^{-1}\|\xi\| \qquad \text{if } \xi \in V^-.$$

Let $\xi^\pm$, $\eta^\pm$ denote the components of ξ, η along $V^\pm$. Then we have

$$\{\xi : \|T^{km}\xi\| \leqslant \frac{1 - \lambda^{-k}}{1 + \lambda^{-k}}\|\xi\|\} \subset \{\xi : \|\xi^+\| \leqslant \lambda^{-k}\|\xi^-\|\}$$

$$\{\eta = T^{km}\xi : \|T^{km}\xi\| \geqslant \frac{1 + \lambda^{-k}}{1 - \lambda^{-k}}\|\xi\|\} \subset \{\eta : \|\eta^-\| \leqslant \lambda^{-k}\|\eta^+\|\}.$$

In the bundle E, we may identify E_y with E_x for y close to x. Thus, if $\xi \in V_y^-$ and

$$\|T_y^{km} - T_x^{km}\| \leq \frac{1 - \lambda^{-k}}{1 + \lambda^{-k}} - \lambda^{-k},$$

we get $\|\xi^+\| \leq \lambda^{-k}\|\xi^-\|$, so that V_y^- depends continuously on y; similarly for V_y^+.

It is now easy to prove (15.1), and since $V^\pm$ are uniquely determined by this, we have $TV^+ = V^+$, $TV^- \subset V^-$.

4. (Hölder continuity of hyperbolic splittings). Let K be a compact hyperbolic set for the map f. We choose a metric d on K such that $c_1\|y - x\| \leq d(x, y) \leq c_2\|y - x\|$ with $0 < c_1 < c_2 < \infty$ when x, y belong to the domain of a local chart of M. Then, differentiability of f implies that, for $k \geq 0$,

$$d(f^k x, f^k y) \leq \sigma^k d(x, y)$$

for some $\sigma > 1$. We shall also use the assumption that the inverse of $f|K$ is Lipschitz, thus:

(1) $$d(f^{-k}x, f^{-k}y) \leq \sigma^k d(x, y).$$

Suppose that f is $C^{\mathbf{r}}$ with $\mathbf{r} > 1$. In particular, f is of class $C^{(1,\alpha)}$ for some $\alpha > 0$. By using a finite number of local charts covering K, we may write

$$\|T_y f^m - T_x f^m\| \leq C'[d(x, y)]^\alpha.$$

Therefore, if $\|Tf^m\| \leq t$, we have

$$\|T_y f^{km} - T_x f^{km}\| \leq t^{k-1}C'[d(x, y)]^\alpha[1 + \sigma^{m\alpha} + \cdots + \sigma^{m\alpha(k-1)}]$$
$$\leq C''[t\sigma^{m\alpha}]^k[d(x, y)]^\alpha.$$

Similarly, if (*) holds,

$$\|T_{f^{-km}x} f^{km} - T_{f^{-km}y} f^{km}\| \leq C''[t\sigma^{m\alpha}]^k[d(x, y)]^\alpha.$$

Use these formulae and those of Problem 3 (where we may take $\lambda = \theta^m/C$, provided this is > 1) to prove Hölder continuity of $x \mapsto V_x^-$ and $x \mapsto V_x^+$ ((1) is assumed to obtain the latter).

If K is a prehyperbolic set, and $K^\dagger = \{(x_k) \in \prod_{k \leq 0} K : f x_{k-1} = x_k\}$ is its hyperbolic cover, a distance is defined on $K^\dagger$ by

$$d^\dagger((x_k),(y_k)) = \sum_{n=0}^{\infty} \sigma^{-n} d(x_{-n}, y_{-n}).$$

Prove the Hölder continuity of $x \mapsto V_x^\pm$ with respect to this distance. (Note that the assumption (1) is not needed here.)

In the case of flows, see the counterexample given by Plante [1].

5. **(Periodic points near a hyperbolic set).** Let K be a compact hyperbolic set for a map f or a semiflow (f^t); we do not assume local product structure. Show that there is a neighborhood U of K, and given $\varepsilon > 0$, there is a $\delta > 0$ such that if $(x_t)_{0 \leq t \leq T}$ is a closed δ-pseudoorbit contained in U (closed means $x_0 = x_T$), then there is a periodic orbit (of period close to T) that ε-shadows (x_t).

The case of a map (which is easier than that of a semiflow) may be treated as follows. Write $X = \{x_0, \ldots, x_{T-1}\}$ and define $h : X \mapsto X$ by $h x_i = x_{i+1}$ for $i = 0, \ldots, T-1$. Also let $i : X \hookrightarrow U$ be the inclusion. Show that the map $k \mapsto f k h^{-1}$ is hyperbolic and has a fixed point j near i; $j X$ is the desired periodic orbit.

[What requires a little thinking is that U, δ may be chosen independently of T. The argument may be extended to a continuous map $i : X \mapsto U$ of a topological space X, a homeomorphism $h : X \mapsto X$, and the study of $k \mapsto g k h^{-1}$ where g is C^1 close to f; see Shub [2], Théorème 7.8.]

6. **(Thickness of Cantor sets in R).**

(a) Given a Cantor set $A \subset \mathbf{R}$ (see Appendix A.3), show that there exists a closed interval $A_0 \subset \mathbf{R}$ and disjoint open intervals $U_0, U_1, U_2, \cdots \subset A_0$ such that

$$A = A_0 \setminus \bigcup_{i=0}^{\infty} U_i.$$

The U_i are called *gaps*; A_0 and the U_i are uniquely determined (but the order of the U_i is of course arbitrary).

(b) With the notation of (a), let

$$A_n = A_0 \setminus \bigcup_{i=0}^{n-1} U_i$$

so that A_n is a union of $n + 1$ disjoint closed intervals, which we call *components*. Then U_n is in one of the components C_n of A_n, and $C_n \backslash U_n = C_n^- \cup C_n^+$ (where C_n^- and C_+^- are components of A_{n+1}). Define

$$\tau(A, (U_i)) = \inf_{n \geqslant 0} \min \left\{ \frac{|C_n^-|}{|U_n|}, \frac{|C_n^+|}{|U_n|} \right\},$$

where $|I|$ denotes the length of the interval I. The *thickness* of A is

$$\tau(A) = \sup\{\tau(A, (U_i)) : \text{ all orderings of } (U_i)\}.$$

(c) Given Cantor sets $A, B \subset \mathbf{R}$ such that neither is contained in a gap of the other, and $\tau(A)\tau(B) > 1$, show that $A \cap B \neq \emptyset$. [This is a consequence of (d) below.]

(d) Under the assumptions of (c), show that if there is a component Γ_n of A_n such that $\Gamma_n \cap B \neq \emptyset$, then there is a component $\Gamma_{n+1} \subset \Gamma_n$ of A_{n+1} such that $\Gamma_{n+1} \cap B \neq \emptyset$.
[We may assume that the gaps U_i of A and V_j of B are so ordered that $\tau(A, (U_i)) \cdot \tau(B, (V_j)) > 1$. With the notation of (b), it suffices to check the case where $\Gamma_n = C_n$ and $\Gamma_{n+1} = C_n^{\pm}$; this is a simple geometric discussion; see Newhouse [3].]

7. (Structurally stable diffeomorphisms of the circle). Let $S^1 = \mathbf{R}/\mathbf{Z}$ be the circle, and $\mathcal{D} = \text{Diff}_+^r(S^1)$ the space of orientation preserving diffeomorphisms of S^1, with the C^r topology, $r \geqslant 1$ (see Section 2.7). Let $H \subset \mathcal{D}$ consist of those f such that $R(f)$ is rational, and all the periodic points of f are hyperbolic.

(a) Let $f \in \mathcal{D}$ and f be rational of the form p/q (p prime to q). Show that $g = f^q$ has a lift $\tilde{g} : \mathbf{R} \mapsto \mathbf{R}$ such that $\tilde{g}x = x$ if and only if x is a periodic point for f.
Write $\psi(x) = \tilde{g}x - x$ and show that x is a hyperbolic periodic point for f if and only if $\psi(x) = 0$ and $\psi'(x) \neq 0$, where ψ' denotes the derivative.

(b) Using the fact that ψ is periodic (and thus defines a function on the compact set $\mathbf{R}/\mathbf{Z}$), show that H is open in $\mathcal{D}$.

(c) Show that $\{f \in \mathcal{D} : R(f) \text{ is rational}\}$ is dense in $\mathcal{D}$. From this, deduce that H is dense in $\mathcal{D}$.

(d) Remember that f is C^r structurally stable if for g sufficiently close to f in $\mathcal{D}$, there is a homeomorphism h of S^1 such that $g = hfh^{-1}$. Show that if $f \in H$, then f is structurally stable.

(e) Show that structural stability of f is equivalent to $f \in H$, so that here the structurally stable diffeomorphisms form an open and dense subset of $\mathcal{D}$.

3 Appendices

Ipsa tunc merui quod nunc plector.

—*Heloysa*

Appendices A, B, and C collect definitions and results from various parts of mathematical analysis for easy reference. Appendix D may be viewed as a continuation of the main text and deals with the specialized question of Axiom-A systems.

A. Sets; Topology, Metric, and Banach Spaces

In this appendix, we review notation, terminology, and some useful results of the theory of sets, topological spaces, metric spaces, and Banach spaces.

A.1. Sets and Maps

We follow the usual set-theoretic terminology. In particular,

$$E \backslash F = \{x \in E : x \notin F\}$$

is the complement of $E \cap F$ in E.

We use interchangeably the words *map, mapping,* and *function* (and to some extent, *set* and *space*); $f : E \mapsto F$ denotes a map f from the set E to the set F, and we also write $f : x \mapsto fx$ when $x \in E$. Instead of fx, we may write $f(x)$. The *composed map* map $g \circ f$ of $f : E \mapsto F$, $g : F \mapsto G$ satisfies $g \circ f(x) = g(f(x))$, and we may write gf instead of $g \circ f$. If $X \subset E$, $f|X$ is the *restriction* of f to X, and $fX = \{fx : x \in X\}$ is the *image* of X by f. If $Y \subset F$, we write $f^{-1}Y = \{x : fx \in Y\}$. If $fx = fy$ implies $x = y$, f is *injective*. If $fE = F$, f is *surjective*. If f is injective and surjective, then f is *bijective* and has an *inverse* $f^{-1} : F \mapsto E$. The *graph* of f is the subset $\{(x, y) : y = fx\}$ of $E \times F$.

A *family* $(x_\alpha)_{\alpha \in A}$ corresponds to a map $\alpha \mapsto x_\alpha$ of the index set A to the set of the x_α. In particular, a family (x_i) indexed by the positive integers is a *sequence*. A family (X_α) of subsets of E is a *covering* of E if $\bigcup_\alpha X_\alpha = E$. The *product* $\prod_\alpha E_a$ of a family $(E_\alpha)_{\alpha \in A}$ of sets E_α is the set of families (x_α), where $x_\alpha \in E_\alpha$ for each $\alpha \in A$. To a family of maps $f_\alpha : E_\alpha \mapsto F_\alpha$, there corresponds a *product map* $\prod_\alpha f_\alpha : (x_a) \mapsto (f_\alpha x_\alpha)$ from $\prod E_\alpha$ to $\prod F_\alpha$. For finite families, i.e., if $A = \{1, \ldots, n\}$, we write

$$\prod E_\alpha = E_1 \times \ldots \times E_n$$
$$(x_\alpha) = (x_1, \ldots, x_n)$$
$$\prod f_\alpha = f_1 \times \ldots \times f_n = (f_1, \ldots, f_n).$$

We denote by E^n the product of n copies of E.

We denote by $\mathbf{Z}$ the ring of integers, by $\mathbf{R}$, $\mathbf{C}$ the real and complex fields. For real $a \leqslant b$, (a, b) is an open interval and $[a, b]$ a closed interval of $\mathbf{R}$.

A.2. Topological Spaces

A set E acquires a *topology*, i.e., becomes a *topological space*, if a set of subsets of E, called *open* subsets, has been chosen. It is required that E and $\emptyset$ be open, and that a finite intersection and arbitrary union of open sets be open. (If all subsets are open, the topology is *discrete*.) The complement of an open set is *closed*. A set S is called a *neighborhood* of $x \in E$ (or of $X \subset E$) if there is an open set Y with $x \in Y \subset S$ (or $X \subset Y \subset S$). If $X \subset E$, there is a largest open set $\subset X$ (the *interior* $\overset{\circ}{X}$ of X) and a smallest closed set $\supset X$ (the *closure* $\overline{X}$ of X). The boundary of X is $\partial X = \overline{X} \backslash \overset{\circ}{X}$. If $\overline{X} = E$ one says that X is *dense* in E. If there is a countable set $\{x_i\}$ dense in E, one says that E is *separable*.

Let (x_i) be a sequence of points of E. A point a is a *limit* of (x_i) if, for every neighborhood U of a, and n positive, there is $i \geq n$ with $x_i \in U$. If n can be chosen such that $x_i \in U$ for all $i \geq n$, then (x_i) is said to be *convergent*, and one writes $x_i \rightarrow a$.

If E, F are topological spaces, a map $f : E \mapsto F$ is *continuous* if $f^{-1}Y$ is open for every open $Y \subset F$. A bijective map that is continuous and has a continuous inverse is called a *homeomorphism*. If $X \subset E$, the intersections with X of the open subsets of E define a topology: the *induced topology*.

The topological space E is *Hausdorff* if, for any distinct $x, y \in E$, there exist disjoint open sets X, Y with $x \in X$, $y \in Y$. In a Hausdorff space, a convergent sequence has a unique limit.

The topological space E is *compact* if, for every covering (X_α) by open sets, there is a finite subfamily (X_{α_i}) that is still a covering of E. Equivalently, one may require the *finite intersection* property: If (F_α) is a family of closed subsets of E with $\bigcap F_\alpha = \emptyset$, one can choose a finite subfamily $(F_{\alpha_1}, \ldots, F_{\alpha_n})$ with $F_{\alpha_1} \cap \cdots \cap F_{\alpha_n} = \emptyset$. If a subset X of the topological space E has compact closure $\overline{X}$, one also says that X is *relatively compact* in E.

The topological space E is *paracompact* if, for every open covering (X_α), there is an open covering (Y_β) that is a *refinement* of (X_α) (i.e., each Y_α is contained in some X_α) and is *locally finite* (i.e., each $x \in E$ has a neighborhood Y that intersects only finitely many Y_α).

To define the *topological product* of a family of topological spaces E_α, consider the subsets $\Pi_\alpha X_\alpha$ where X_α is open in E_α for each α, and $X_\alpha \neq E_\alpha$ for finitely many values of the index α only. The *product topology* on $\Pi_\alpha E_\alpha$ is the topology for which the open sets are the above $\Pi_\alpha X_\alpha$ and their unions. A topological product of compact spaces is compact.

The topological space E is *connected* if an open and closed subset X of E is necessarily $\emptyset$ or E.

For each finite covering (X_α) of the topological space E by open sets, suppose that we can find a new covering (Y_α), with open $Y_\alpha \subset X_\alpha$, such that every intersection of $n + 2$ sets Y_α is empty. We write then $\dim E \leq n$. If we have $\dim E \leq n$ but not $\dim E \leq n - 1$, we say that E has *topological dimension* n.

A.3. Metric Spaces

A *metric* on a set E is defined by a *distance* function $d : E \times E \mapsto \mathbf{R}$ that satisfies $d(x, x) = 0$, $d(x, y) = d(y, x) > 0$ if $x \neq y$, and $d(x, y) + d(y, z) \geqslant d(x, z)$ (*triangle inequality*). For instance, the *Euclidean metric* on $\mathbf{R}^n$ is defined by $d((x_i), (y_i)) = (\sum_{i=1}^{n} (y_i - x_i)^2)^{1/2}$. A space E with metric is a *metric space*. The *diameter* of E is the sup of the distances between pairs of points, and E is said to be *bounded* if it has finite diameter. If $X \subset E$, the restriction of d to $X \times X$ is a metric: the *induced metric*.

Given $a \in E$ and $R > 0$, we call $B_a(R) = \{x \in E : d(a, x) < R\}$ the *open ball* of center a and radius R. We define a set $X \subset E$ to be open if for each $x \in X$, there is ε such that $B_x(\varepsilon) \subset X$. These open sets define a topology on E, and one can show that this topology is Hausdorff and paracompact.

If a topological space E has a metric that defines the topology as above, the metric is said to be *compatible* with the topology, and E is said to be *metrizable*. In a compact metrizable space, every sequence has a convergent subsequence.

A sequence (x_i) converges to a in a metric space E if and only if $\lim_{i \to \infty} d(x_i, a) = 0$. If a sequence (x_i) satisfies $\lim_{i,j \to \infty} d(x_i, x_j) = 0$, then (x_i) is a *Cauchy sequence*. If E is such that every Cauchy sequence is convergent, we say that E is *complete*. A compact metric space is always complete. In a complete metric space, a set containing a countable intersection of dense open subsets is called *residual*; a residual set is dense (*Baire's theorem*).

A map f of a metric space E to itself is a *contraction* if there is $\alpha < 1$ such that $d(fx, fy) \leqslant \alpha d(x, y)$ for all $x, y \in E$. If E is complete then the contraction f has a unique fixed point a, and $f^i x \to a$ for all $x \in E$ (*contraction mapping theorem*).

A compact metrizable space E is called a *Cantor space* if the following conditions are both satisfied

   (a) E is completely disconnected (i.e., for any two distinct $x, y \in E$, there are disjoint open sets $X \ni x$, $Y \ni y$, such that $X \cup Y = E$),
   (b) E contains no isolated points (i.e., if $x \in E$, $E \backslash \{x\}$ cannot be closed).

Every two Cantor spaces are homeomorphic.[1] The topological product

---

[1] See, for instance, Moise [1], Chapter 12.

$\prod A_i$ of a countable infinite family of finite sets A_i with the discrete topology and card $A_i > 1$ is a Cantor space. The topological dimension of a Cantor space is 0.

Given a compact metric space $E \neq \emptyset$, and a real number $r > 0$, we denote by $\sigma = (\sigma_k)$ a countable covering of E by sets σ_k with diameters $d_k = \operatorname{diam} \sigma_k \leqslant r$. For every $\alpha \geqslant 0$, we write

$$m_r^\alpha(E) = \inf_\sigma \sum_k (d_k)^\alpha.$$

When r decreases to 0, $m_r^\alpha(E)$ increases to a (possibly infinite) limit $m^\alpha(E)$ called the *Hausdorff measure of E in dimension α*. We write

$$\dim_H E = \sup\{\alpha : m^\alpha(E) > 0\}$$

and call this quantity the *Hausdorff dimension* of E. Note that $m^\alpha(E) = +\infty$ for $\alpha < \dim_H E$ and $m^\alpha(E) = 0$ for $\alpha > \dim_H E$.

A.4. Banach Spaces, Hilbert Spaces

In this section and the next one, we shall discuss both *real* and *complex* Banach spaces. In the rest of the monograph, Banach space means real Banach space, unless otherwise indicated.

Let E be a vector space over $\mathbf{R}$ (or $\mathbf{C}$). A *norm* $\|\cdot\|$ is a function $E \mapsto \mathbf{R}$ such that $\|0\| = 0$, $\|x\| > 0$ if $x \neq 0$, $\|\alpha x\| = |\alpha| \cdot \|x\|$ is $\alpha \in \mathbf{R}$ (or $\mathbf{C}$) and $\|x + y\| \leqslant \|x\| + \|y\|$. Associated with $\|\cdot\|$, there is a metric

$$d(x, y) = \|y - x\|.$$

If E is complete with respect to this metric, one says that E, equipped with the norm $\|\cdot\|$, is a real (or complex) *Banach space*.

A norm $\|\cdot\|'$ is said to be *equivalent* to $\|\cdot\|$ if

$$\alpha^{-1}\|\cdot\| \leqslant \|\cdot\|' \leqslant \alpha\|\cdot\|$$

for some $\alpha \geqslant 1$. Equivalent norms define the same topology. Conversely, all norms defining this topology (*permissible norms*) are equivalent. On $\mathbf{R}^m$, every norm is equivalent to the Euclidean norm: $\|x\| = (\sum_1^m x_i^2)^{1/2}$. The open ball $B_a(R)$ in E will also be denoted by $E_a(R)$.

A real Banach space E extends to a complex Banach space $E_{\mathbf{C}} = \{x + iy : x, y \in E\}$ with norm $\|x + iy\| = (\|x\|^2 + \|y\|^2)^{1/2}$; $E_{\mathbf{C}}$

is called the *complexification* of E. A complex Banach space may be considered as a real Banach space by restricting the scalars to $\mathbf{R}$. In this manner, E may be considered as a subspace of $E_{\mathbf{C}}$.

A closed subspace of a Banach space is again a Banach space for the induced norm. The *product* $E \times F$ of two Banach spaces E, F is a Banach space with respect to the norm

(A.1) $\|(x, y)\| = \max\{\|x\|, \|y\|\}.$

It is convenient to define the *direct sum* $E \oplus F$ as being $E \times F$ with the identifications $E \mapsto E \times \{0\}$, $F \mapsto \{0\} \times F$. Two closed subspaces E, F of a Banach space G are *complementary* if $E \cap F = \{0\}$ and $E + F = G$. In that case, $x \oplus y \mapsto x + y$ identifies $E \oplus F$ with G, and the norm (A.1) is equivalent to the original norm of G. If G is a Banach space and E a closed subspace, each one of the following conditions is sufficient for the existence of a closed complement F:

(a) E is finite dimensional,
(b) E is finite codimensional (i.e., there exists $\widetilde{F}$ of finite dimension such that $E + \widetilde{F} = G$),
(c) G is a Hilbert space (see below).

Let F be a closed subspace of the Banach space E. The relation $x - y \in F$ between $x, y \in E$ is an equivalence relation. The equivalence classes $[x]$ form a linear space that is a Banach space with respect to the norm

$$\|[x]\| = \inf_{x \in [x]} \|x\|.$$

This is the *quotient Banach space* E/F. Clearly $E \oplus F/F$ can be identified with E.

The real Banach space E is a *Hilbert space* if there is a bilinear map $x, y \mapsto (x, y)$ of $E \times E$ to $\mathbf{R}$ (the *scalar product*) such that $(x, y) = (y, x)$ and $(x, x) = \|x\|^2$; similarly, for a complex Hilbert space. But in the latter case, $x, y \mapsto (x, y)$ maps $E \times E$ to $\mathbf{C}$, is $\mathbf{C}$-linear in y, and antilinear in x; here $(x, y) = (y, x)^*$ and again $(x, x) = \|x\|^2$. A Banach space is a Hilbert space if and only if $\|x + y\|^2 + \|x - y\|^2 = 2\|x\|^2 + 2\|y\|^2$.

A.5. Operators in Banach Spaces

Let E, F be real, resp. complex, Banach spaces. If $A : E \mapsto F$ is a linear map (i.e., $\mathbf{R}$-linear, resp. $\mathbf{C}$-linear), let

(A.2) $\|A\| = \sup_{x : \|x\| \leqslant 1} \|Ax\|.$

Then A is continuous if and only if it is bounded, i.e., if $\|A\| < \infty$. We denote by $\mathcal{L}(E; F)$ the Banach space of (bounded) linear maps $A : E \mapsto F$ with the norm (A.2). We also write

$$m(A) = \inf_{x:\|x\|=1} \|Ax\|.$$

A map $A \in \mathcal{L}(E; F)$ has an inverse $A^{-1} \in \mathcal{L}(F; E)$ if and only if it is bijective (and then $m(A) = \|A^{-1}\|^{-1}$).

We write $\mathcal{L}(E; E) = \mathcal{L}(E)$. This is an algebra with, as unit element, the identity map $\mathbf{1}$ on E. The elements of $\mathcal{L}(E)$ are called *linear operators* on E.

We say that a multilinear map $B : E^k \mapsto F$ is bounded if $\|B\| < \infty$, where

$$\|B\| = \sup\{\|B(x_1, \dots, x_k)\| : \|x_1\| \leqslant 1, \dots, \|x_k\| \leqslant 1\}.$$

Such maps form a Banach space $\mathcal{L}^k(E; F)$.

If E, F are real Banach spaces, a linear map $A : E \mapsto F$ extends uniquely to a C-linear map $E_{\mathbf{C}} \mapsto F_{\mathbf{C}}$ with the same norm, this map is again denoted by A (it is called a *real* operator).

Let E be a complex Banach space, and $A \in \mathcal{L}(E)$. The *spectrum* of A is the compact subset of $\mathbf{C}$ formed by those λ such that $A - \lambda\mathbf{1}$ is not invertible in $\mathcal{L}(E)$. If E is a real Banach space, and $A \in \mathcal{L}(E)$, we may identify A with an element of $\mathcal{L}(E_{\mathbf{C}})$, and the spectrum of that element is called the spectrum of A. The linear operator A is said to be *hyperbolic* if its spectrum is disjoint from the unit circle $\{z \in \mathbf{C} : |z| = 1\}$, and *$\rho$-pseudohyperbolic* if it is disjoint from the circle $\{z \in \mathbf{C} : |z| = \rho\}$.

The operators A with spectrum disjoint from a closed set $\Gamma \subset \mathbf{C}$ form an open set in the Banach space $\mathcal{L}(E)$. If Γ is a closed contour, disjoint from the spectrum, and S is the part of the spectrum enclosed in Γ, the operator

$$P = \frac{1}{2\pi i} \oint_\Gamma \frac{dz}{z\mathbf{1} - A}$$

is called the *projection associated* with Γ. It satisfies $P^2 = P, PA = AP$, and depends on A continuously (in fact, smoothly). If Γ surrounds the whole spectrum, then $P = \mathbf{1}$. If A is a real operator, and Γ is symmetric with respect to the real axis, then P is a real operator. The closed subspaces PE and $(\mathbf{1} - P)$ of E are complementary.

The *spectral radius* of A is the number

$$\max\{|z| : z \in \text{ spectrum of } A\} = \lim_{n \to \infty} \|A^n\|^{1/n}.$$

It is *not* a norm on $\mathcal{L}(E)$.

A.6. Note

Theoretical and mathematical physicists are faced with the recurrent problem of rapidly acquiring a working knowledge of exotic fields of mathematics. Although there is no easy general solution to this problem, the *Methods of Modern Mathematical Physics* by Reed and Simon [1], and the *Encyclopedic Dictionary of Mathematics* by the Nihon Sugakkai (Mathematical Society of Japan) [1] are quite helpful in different ways. I also have a personal liking for Bourbaki's *Fascicules de Résultats* [1], [2], [3], [5].

B. Manifolds

This is a review of definitions and results on differentiability and differentiable manifolds in infinite dimension (Banach manifolds).

B.1. Differentiability

From now on, Banach spaces will be real Banach spaces unless otherwise indicated.

Let E, F be Banach spaces, and U an open subset of E. We say that the function $f : U \mapsto F$ is (Fréchet) *differentiable* at $x \in U$, and that its *derivative at* x is $D_x f \in \mathcal{L}(E, F)$ if

$$\lim_{\xi \to 0, \xi \neq 0} \frac{\|f(x + \xi) - f(x) - (D_x f)\xi\|}{\|\xi\|} = 0.$$

We also write $D_x f = f'(x)$. If $f'(x)$ is defined for all $x \in U$, then $f' : U \mapsto \mathcal{L}(E; F)$ is called the *derivative* of f in U. If f is continuously differentiable, i.e., if f' is continuous, we have

$$\lim_{\xi, \eta \to 0, \xi \neq \eta} \frac{\|f(x + \xi) - f(x + \eta) - (D_x f)(\xi - \eta)\|}{\|\xi - \eta\|} = 0.$$

If it exists, the kth derivative at x is defined recursively by $D_x^0 f = f(x)$, and $D_x^k f = D_x D_x^{k-1} f \in \mathcal{L}(E^k; F)$. We also write $D_x^k f = f^{(k)}(x)$. If the derivatives $f^{(k)} : U \mapsto \mathcal{L}^k(E; F)$ are continuous for $k = 0, \dots, r$, we say that f is C^r, or of class C^r, in U; the rth derivative $D_x^r f$ at $x \in U$ is then a symmetric multilinear map $E^r \mapsto F$.

We say that a function $g : U \mapsto F$ is *Hölder* continuous of exponent α (with $0 < \alpha \leqslant 1$) if each $x \in U$ has a neighborhood V such that

$$\text{(B.1)} \quad |g|_{V,a} = \sup \left\{ \frac{\|g(y) - g(z)\|}{\|y - z\|^\alpha} : y, z \in V, \quad y \neq z \right\} < \infty.$$

In particular, if $\alpha = 1$, then g is *Lipschitz* and $|g|_{V,1}$ is the *Lipschitz constant* of g in V. If $f^{(r)} : U \mapsto \mathcal{L}^r(E, F)$ is Hölder continuous of exponent α, we say that f is of class $C^{(r,\alpha)}$.

A function f is of class C^∞ if each x has a neighborhood V on which every derivative exists and is bounded. If furthermore, for each $a \in U$, there exist $A, R > 0$ such that

$$\|f^{(k)}(a)\| \leqslant k! A R^{-k},$$

then f is of class C^ω, i.e., *real analytic*, and we have a convergent Taylor expansion

$$f(x) = \sum_{k=0}^{\infty} \tfrac{1}{k!} D_a^k f(x - a, \dots, x - a)$$

for $|x - a| < R$. There is thus an extension of f to a holomorphic function $\widetilde{U} \mapsto F_{\mathbf{C}}$, where $\widetilde{U}$ is a neighborhood of U in the complexified Banach space $E_{\mathbf{C}}$. [A function $\tilde{f}$ from an open set $\widetilde{U}$ of a complex Banach space $\widetilde{E}$ to a complex Banach space $\widetilde{F}$ is *holomorphic* if it is locally given by convergent Taylor expansions. Equivalently, $\tilde{f}$ is holomorphic if, for each $x \in \widetilde{U}$, the derivative $D_x \tilde{f}$ exists as a (**C**-linear) bounded map $\widetilde{E} \mapsto \widetilde{F}$. The C^ω functions on subsets of a real Banach space are thus restrictions of **C**-differentiable functions on subsets of a complex Banach space.[2]]

It is convenient to say that the continuous functions (resp. the Hölder-continuous functions of exponent α) are of class C^0 (resp. $C^{(0,\alpha)}$). A total order is defined on the symbols $r, (r, \alpha), \infty, \omega$ such that

$$r < (r, \alpha) < r + 1 < \infty < \omega,$$

and we write $|r| = r$, $|(r, \alpha)| = r + \alpha$, $|\infty| = \infty$, $|\omega| = \omega$. If $\mathbf{r}$ stands for any of the above symbols $C^{\mathbf{r}}(U, F)$ denotes the space of functions $f : U \mapsto F$ of class $C^{\mathbf{r}}$. Then $\mathbf{s} < \mathbf{r}$ implies $C^{\mathbf{s}} \supset C^{\mathbf{r}}$. A function of class $C^{\mathbf{r}}$ is *differentiable* (or *smooth*) if $\mathbf{r} \geqslant 1$.

If $\mathbf{r} \geqslant 1$, the derivative of a function of class $C^{\mathbf{r}}$ is of class $C^{\mathbf{r}-1}$ (where $(r, \alpha) - 1 = (r - 1, \alpha)$, $\infty - 1 = \infty$, $\omega - 1 = \omega$). Composition of two functions of class $C^{\mathbf{r}}$ is of class $C^{\mathbf{r}}$ if $\mathbf{r} \geqslant 1$ (or $\mathbf{r} = 0$, $\mathbf{r} = (0, 1)$, but $C^{0,\alpha} \circ C^{0,\beta} \subset C^{0,\alpha\beta}$ only). It is inconvenient to define a topology on

---

[2] See Bourbaki [5].

the spaces $C^{\mathbf{r}}(U, F)$. Instead, we introduce smaller spaces, appropriate for local studies. Using (A.1), let

$$\|f\|_{U,0} = \sup\{\|f(x)\| : x \in U\},$$
$$\|f\|_{U,r} = \max\{\|f^{(k)}\|_{U,0} : k = 0, 1, \ldots, r\},$$
$$\|f\|_{U,(r,\alpha)} = \max\{\|f\|_{U,r}, |f^{(r)}|_{U,\alpha}\}.$$

We define $C^{\mathbf{r}}(U, F) \subset C^{\mathbf{r}}(U, F)$ for $\mathbf{r} < \infty$ to be the Banach space of elements with finite norm $\|\cdot\|_{U,\mathbf{r}}$. We also write $C^{\infty}(U, F) = \bigcap_r C^r(U, F)$ and call $\mathcal{O}$ an open set if, for every $f \in \mathcal{O}$, there are $r \geqslant 0$ and $\varepsilon > 0$ such that $\{g : \|g - f\|_{U,r} < \varepsilon\} \subset \mathcal{O}$. This defines a topology on $C^{\infty}(U, F)$, and there is a metric, compatible with this topology, for which $C^{\infty}(U, F)$ is complete (it is a Fréchet space). To handle real analytic functions, we choose an open set $\widetilde{U} \subset E_{\mathbf{C}}$ with $\widetilde{U} \cap E \supset U$, and consider the Banach space $\mathcal{H}(\widetilde{U}, F_{\mathbf{C}})$ of holomorphic maps $\widetilde{U} \mapsto F_{\mathbf{C}}$, with the norm $\|\cdot\|_{\widetilde{U},0}$. In this way, we can define continuity with respect to a $C^{\mathbf{r}}$ function for any $\mathbf{r}$.

We say that the Banach space E has the $C^{\mathbf{r}}$ *extension property*, $\mathbf{r} \leqslant \infty$, if there exists a function $\varphi \in C^{\mathbf{r}}(E, \mathbf{R})$ vanishing outside of the unit ball $E_0(1)$ and equal to 1 in $E_0(\delta)$ for some $\delta > 0$. The space E has the $C^{\mathbf{r}}$ extension property as soon as there exists a nontrivial function in $\mathcal{C}^{\mathbf{r}}(\mathcal{E}, \mathbf{R})$ with bounded support, for instance if the norm $x \mapsto \|x\|$ is $C^{\mathbf{r}}$ outside of the origin. If E is finite dimensional or is a Hilbert space, the C^{∞} extension property holds.[3]

Replacing the norms on Banach spaces by equivalent norms does not change the definition of derivatives, differentiability classes, or the $C^{\mathbf{r}}$ extension property.

B.2. Compact Maps, Leray–Schauder Degree[4]

Let E, F be Banach spaces, and $U \subset E$. The continuous map $f : U \mapsto F$ is said to be *compact* if, for bounded $B \subset U$, the image fB has compact closure in F. If B is closed, convex, bounded, f compact, and $fB \subset B$, then f has a fixed point in B (*Schauder fixed-point theorem*).

---

[3]For further discussion, see Bonic and Frampton [1]. The definition given here is (in principle) more restrictive than that of Bonic and Frampton, because we assume $\varphi \in C^{\mathbf{r}}$ instead of $\varphi \in C^{\mathbf{r}}$. This strengthening is used in the proof of Lemma 14.5 (for $\mathbf{r} = 1$).

[4]See Berger [1].

If $A : E \mapsto E$ is linear, invertible, and if $A - 1$ is compact, we define

$$\deg A = (-1)^{\alpha},$$

where α is the number of negative eigenvalues of A. In particular, if E is finite dimensional, $\deg A$ is the sign of $\det A$ ($+1$ if A preserves orientation, -1 if A reverses orientation).

Let U be a bounded open subset of E, with boundary $\partial U = \overline{U} \backslash U$. We assume that $f : \overline{U} \mapsto E$ differs from the identity by a compact map, and that $a \in E \backslash f(\partial U)$. The *Leray–Schauder degree* $\deg(f, a, U) \in \mathbf{Z}$ is then defined and has the following properties.

(a) $\deg(f, a, U)$ *depends continuously on* f *(with respect to* $\| \cdot \|_{U,0}$*) and a.*

(b) *If the equation* $f(x) = a$ *has a finite number of solutions* x_i *in* U, *and if* f *is differentiable at* x_i *with* $D_{x_i} f$ *invertible and* $D_{x_i} f = \mathbf{1} +$ *compact operator, then*

(B.1) $$\deg(f, a, U) = \sum_i \deg(D_{x_i} f).$$

Property (a) implies that $\deg(f, a, U)$ is constant when f and a change continuously, respecting $a \notin f(\partial U)$. In particular, one can find approximations of f such that (B.1) applies, and $\deg$ is thus uniquely defined. If E is finite dimensional, $\deg(f, a, U)$ is called the *Brower degree*.

To prove that in a compact map, $g : \overline{U} \mapsto E$ has a fixed point in U, it suffices to show that $\deg(\mathbf{1} - g, 0, U) \neq 0$, which is often possible by continuous deformation of g (homotopy), using Property (a).

B.3. Inverse Function Theorem and Consequences ($r \geqslant 1$).

B.3.1. Theorem (inverse function theorem).[5] *Let* E, F *be Banach spaces,* U *an open subset of* E *and* $a \in U$. *We assume that* $f \in C^{\mathbf{r}}(U, F)$ *with* $\mathbf{r} \geqslant 1$ *(*$\mathbf{r}$ *is an integer* r, *or* (r, α), *or* ∞, *or* ω*) and that* $D_a f$ *has an inverse* $(D_a f)^{-1} \in \mathcal{L}(F; E)$. *Then there is an open set* $\widetilde{U}$ *such that* $a \in \widetilde{U} \subset U$,

---

[5] The C^r case, r integer, is proved in Lang [1]. The other cases follow readily. (See also Bourbaki [5].)

and $f|\widetilde{U}$ has a unique inverse $g : f(\widetilde{U}) \mapsto \widetilde{U}$. The function g is $C^{\mathbf{r}}$ and $D_{f(y)}g = (D_y f)^{-1}$ for $y \in \widetilde{U}$.

This result is basic for the theory of differentiable manifolds and is the reason for the special role played by Banach spaces in this theory. (The inverse function theorem does not extend in a simple usable manner to more general spaces).

B.3.2. Corollary. *If $f \in C^{\mathbf{r}}(U, F)$ and f has a C^1 inverse g, then g is of class $C^{\mathbf{r}}$.*

Note however that the C^ω map $x \mapsto x^3$ of $\mathbf{R}$ to itself has a C^0 inverse $y \mapsto y^{1/3}$, which is not differentiable.

B.3.3. Corollary (Implicit function theorem). *Let E, F be Banach spaces, U (resp. V) an open set in E (resp. F), and $a \in U$, $b \in V$. Let $f \in C^{\mathbf{r}}(U \times V, E)$, and the derivative $D^1_{(a,b)}f$ with respect to the first argument be invertible in $\mathcal{L}(E)$. There are then open sets $\widetilde{U}$, $\widetilde{V}$ such that $a \in \widetilde{U} \subset U$, $b \in \widetilde{V} \subset V$ and a unique function $g : \widetilde{V} \mapsto \widetilde{U}$ such that $f(g(y), y) = f(a, b)$ for $y \in \widetilde{V}$. The function g is $C^{\mathbf{r}}$ and $D_y g = -(D^1_{(g(y),y)}f)^{-1} D^2_{(g(y),y)}f$ (where D^2 denotes the derivative with respect to the second argument).*

This follows from Theorem B.3.1 by considering the map $(x, y) \mapsto (f(x, y), y)$ of $U \times V$ to $E \times F$.

Let U, V be open subsets of the Banach spaces E, F. If $f \in C^{\mathbf{r}}(U, F)$ maps U to V and has a smooth inverse $f^{-1} : V \mapsto U$, then f is called a $C^{\mathbf{r}}$ *diffeomorphism*.

Let U be an open subset of the Banach space E. A subset X of U is a $C^{\mathbf{r}}$ *submanifold* if for each $x \in X$, there is an open set $\widetilde{U}$ with $x \in \widetilde{U} \subset U$, and a $C^{\mathbf{r}}$ diffeomorphism $f : \widetilde{U} \mapsto V \times W \subset F \times G$, such that $f(X \cap \widetilde{U}) = V \times \{0\}$. ($V$, W are open subsets of the Banach spaces F, G, and $0 \in W$). We define $\mathrm{codim}_x X = \dim G$; if independent of x, it is the *codimension* of X. The *tangent space* $T_x X$ to X at x is the image of V by $(D_x f)^{-1}$. [C^0 or $C^{0,1}$ submanifolds can be defined similarly, with f and f^{-1} of class C^0 or $C^{0,1}$; such manifolds do not have tangent spaces.]

B.3.4. Corollary (on immersions). *Let $U \ni a$, $V \ni b$ be open subsets of the Banach spaces E, F, and $fa = b$, where $f \in C^{\mathbf{r}}(U, F)$. If $D_a f$ is injective and if $(D_a f)E$ is a closed subspace of F which has a closed complement, one says that f is an* immersion *at a. There is then an open set $\widetilde{U} \ni a$ such that $f\widetilde{U}$ is a $C^{\mathbf{r}}$ submanifold of V tangent to $(D_a f)E$ at b.*

B.3.5. Corollary (on transversality). *Let $U \ni a$, $V \ni b$ be open subsets of the Banach spaces E, F, and $fa = b$, where $f \in C^{\mathbf{r}}(U, F)$. Let $Y \ni b$ be a submanifold of V. If $(D_a f)E + T_b Y = F$, and $(D_a f)^{-1} T_b Y$ has a closed complement in E, one says that f is* transversal *to Y at a. There is then as open set $\widetilde{U} \ni a$ such that $\widetilde{U} \cap f^{-1} Y$ is a $C^{\mathbf{r}}$ submanifold of U tangent to $(D_a f)^{-1} T_b Y$ at a.*

For completeness, we state a result which is dual to the corollary on immersions.

B.3.6. Corollary (on submersions). *Let $U \ni a$, $V \ni b$ be open subsets of the Banach spaces E, F, and $fa = b$, where $f \in C^{\mathbf{r}}(U, F)$. If $D_a f$ is surjective and if $(D_a f)^{-1}\{0\}$ has a closed complement in E, one says that f is a* submersion *at a. There are then open sets $\widetilde{U}$, $\widetilde{V}$, with $\widetilde{U} \ni a$, $\widetilde{V} \supset f\widetilde{U}$, and diffeomorphisms $\varphi : \widetilde{U} \mapsto U_1 \times U_2 \subset E_1 \times E_2$ and $\psi : \widetilde{V} \mapsto V_1 \subset E_1$ such that $\psi \circ f \circ \varphi^{-1}(x_1, x_2) = x_1$.*

B.4. Manifolds ($\mathbf{r} \geqslant 1$, or $\mathbf{r} = 0$, or $\mathbf{r} = (0, 1)$).

We have defined differentiable maps and submanifolds for open subsets of Banach spaces. These concepts extend to the more general setting of *Banach manifolds*, which locally have the differentiable structure of a Banach space.

Let M be a Hausdorff topological space.[6] A *chart* of M is a triplet (U, φ, E), where U is an open subset of M and φ a homeomorphism of U to an open subset of the Banach space E. Let $\mathbf{r} \geqslant 1$, or $\mathbf{r} = 0$, or $\mathbf{r} = (0, 1)$. A $C^{\mathbf{r}}$ *atlas* is a set of charts $(U_\alpha, \varphi_\alpha, E_\alpha)$ such that the U_α cover M and $\varphi_\beta \circ \varphi_\alpha^{-1}$ is of class $C^{\mathbf{r}}$ where defined. Two atlases are called $C^{\mathbf{r}}$ equivalent if their union is a $C^{\mathbf{r}}$ atlas. The set M together with a $C^{\mathbf{r}}$ equivalence class of atlases is called a $C^{\mathbf{r}}$ Banach manifold, or simply $C^{\mathbf{r}}$ *manifold*. A chart (U, φ, E) of one of the atlases of M

---

[6] One can study manifolds that are not Hausdorff, but usually one imposes both the Hausdorff condition and paracompactness, or even separability (see below).

is called a C^r chart of M; if $U \ni x$, it is a *chart at x*. If $r \geqslant 1$, M is a *differentiable* (or *smooth*) manifold. Obviously, a C^r manifold may also be considered as a C^s manifold for $s < r$. If the C^r manifold M has an atlas $\{(U_\alpha, \varphi_\alpha, E_\alpha)\}$ where all E_α are equal to a Banach space E, M is called *of type E* and $\dim M = \dim E$ is the *dimension* of M. In particular, a manifold of type $\mathbf{R}^m$ is *finite dimensional* (and more precisely *m-dimensional*).

An open subset U of a Banach space E is a C^ω manifold with the atlas constituted of the only chart (U, φ, E), where $\varphi : U \mapsto E$ is the canonical map.

B.4.1. Proposition.[7] *The implications* (a) $\Rightarrow$ (b) $\Leftrightarrow$ (c) $\Leftrightarrow$ (d) *hold between the following conditions on the topology of a manifold M,*

(a) *M is separable,*
(b) *M is paracompact,*
(c) *M is homeomorphic to a complete metric space,*
(d) *M is metrizable.*

The various terms are defined in Appendix A.2 and A.3.

Let M, N be C^r manifolds and $s \leqslant r$. We say that a continuous map $f : M \mapsto N$ is of class C^s if $\psi_\beta \circ f \circ \varphi_\alpha^{-1}$ is of class C^s, where defined, for charts $(U_\alpha, \varphi_\alpha, E_\alpha)$ and $(V_\beta, \psi_\beta, F_\beta)$ of C^r atlases of M, N. We denote by $C^s(M, N)$ the space of such maps. If $s \geqslant 1$, f is *differentiable* (or *smooth*), and if f has a smooth inverse, then f is called a C^s *diffeomorphism*. If M is compact, the diffeomorphisms $M \mapsto N$ form an open subset of $C^s(M, N)$.

B.4.2. Proposition.[8] *Let M be a separable C^r manifold of type E, where E has the C^r extension property. For every locally finite open covering (X_α) of M, there is a family (ψ_α) of C^r functions $M \mapsto \mathbf{R}$ such that $\psi_\alpha(x) \geqslant 0$, $\psi_\alpha(x) = 0$ for $x \notin X_\alpha$, and $\sum_\alpha \psi_\alpha = 1$ on M. Such a family is called a C^r partition of unity.*

By using a smooth partition of unity, it is possible to approximate continuous functions by smooth ones (*regularization*).

---

[7]The implication (a)$\Rightarrow$(b) is in Bonic and Frampton [1], (b)$\Rightarrow$(c) in Bourbaki [5]; (c)$\Rightarrow$(d) is trivial, and (d)$\Rightarrow$(b) in J. L. Kelley [1].

[8]See Bonic and Frampton [1].

If M, N are manifolds with $C^{\mathbf{r}}$ atlases $\{(U_\alpha, \varphi_\alpha, E_\alpha)\}$, $\{(V_\beta, \psi_\beta, F_\beta)\}$, the *product* $M \times N$ with the atlas $\{(U_\alpha \times V_\beta, \varphi_\alpha \times \psi_\beta, E_\alpha \times F_\beta)\}$ is a $C^{\mathbf{r}}$ manifold.

A subset X of a manifold M is a $C^{\mathbf{r}}$ *submanifold* if, for every chart (U, φ, E) of a $C^{\mathbf{r}}$ atlas of M, $\varphi(X \cap U)$ is a $C^{\mathbf{r}}$ submanifold of $\varphi(U)$; X has then a natural $C^{\mathbf{r}}$ manifold structure. If the codimension of $\varphi(X \cap U)$ is constant when $X \cap U \neq \emptyset$, this is called the *codimension* of $X \neq \emptyset$. Each open subset of M is a $C^{\mathbf{r}}$ submanifold of codimension 0. Every submanifold X of M is a *closed submanifold* of some open subset of M.

Suppose now $\mathbf{r} \geqslant 1$. By using charts, one extends, in the obvious way, to manifolds the definitions of immersion at a point (Corollary B.3.4) and transversality (Corollary B.3.5). One says that $f \in C^{\mathbf{r}}(M, N)$ is an *immersion* if it is an immersion at each $x \in M$. One says that $f \in C^{\mathbf{r}}(M, N)$ is *transversal* to a $C^{\mathbf{r}}$ submanifold Y of N if it is transversal at each $x \in f^{-1}Y$. Two submanifolds X, Y of M are transversal at $x \in M \cap N$ (resp. transversal) if the inclusion $X \mapsto M$ is transversal to Y at x (resp. transversal to Y).

B.4.3. Proposition (on embeddings). *If $r \geqslant 1$ and $f \in C^{\mathbf{r}}(M, N)$ is an immersion, and if f defines a homeomorphism of M to $f(M)$ (with the induced topology), then $f(M)$ is a $C^{\mathbf{r}}$ submanifold of N, and f defines a $C^{\mathbf{r}}$ diffeomorphism of M to $f(M)$. One then says that f is an* embedding *of M into N.*

Note that an immersion may be injective without being an embedding. [For instance, the map $\mathbf{R}^2 \mapsto T^2$ defined by $t \mapsto (t(\mathrm{mod}1), \alpha t(\mathrm{mod}1))$ with α irrational is an immersion with dense image.]

B.4.4. Proposition. *If $\mathbf{r} \geqslant 1$ and $f \in C^{\mathbf{r}}(M, N)$ is transversal to the $C^{\mathbf{r}}$ submanifold Y of N, then $f^{-1}Y$ is a $C^{\mathbf{r}}$ submanifold of M.*

B.4.5. Theorem (Whitney's theorem).[9] *If M is an m-dimensional separable $C^{\mathbf{r}}$ manifold, $\mathbf{r} \geqslant 1$, there is a $C^{\mathbf{r}}$ embedding $f : M \mapsto \mathbf{R}^{2m+1}$ such that fM is a closed C^ω submanifold of $\mathbf{R}^{2m+1}$.*

If $1 \leqslant \mathbf{r} \leqslant \mathbf{s}$, any separable m-dimensional $C^{\mathbf{r}}$ manifold M is thus $C^{\mathbf{r}}$ diffeomorphic to a $C^{\mathbf{s}}$ manifold $\widetilde{M}$. One can show that any two such

[9]See De Rham [1], Section 3, or Hirsch [2], Section 1.3.

manifolds $\widetilde{M}$ are C^s diffeomorphic. On the other hand, two smooth manifolds may be homeomorphic without being diffeomorphic. For instance, there are spheres of dimension $m \geq 7$ (exotic spheres) that are homeomorphic but not diffeomorphic to the usual sphere S^m.[10]

B.4.6. Theorem (Thom's transversality lemma).[11] *Let M, N, Y be C^r manifolds, where M is compact and Y a closed submanifold of N. The set of maps $M \mapsto N$ that are transversal to Y is open and dense in $C^r(M, N)$ when $r \geq 1$.*

(To define the topology of a neighborhood of f in $C^r(M, N)$, we cover M by finite collections of open sets U_α and, by using charts of N, identify $f|U_\alpha$ with a map $U_\alpha \mapsto F_\alpha$. If $\mathcal{N}_\alpha$ is a neighborhood of $f|U_\alpha$ in $C^r(U_\alpha, F_\alpha)$, then $\{g : g|U_\alpha \in \mathcal{N}_\alpha\}$ defines a neighborhood of f in $C^r(M, N)$.)

Note that in the theory of manifolds, the norms of Banach spaces need be defined only up to equivalence.

One often does not make explicit the class C^r of manifolds, diffeomorphisms, etc., when this is clear from context.

B.5 Tangent Bundle and Vector Fields

Let $\{(U_\alpha, \varphi_\alpha, E_\alpha)\}$ be an atlas for a C^r manifold M, with $r \geq 1$. In the union $\bigcup_\alpha (\varphi_\alpha U_\alpha) \times E_\alpha$, we identify the points $(\varphi_\alpha x, Y)$ and $(\varphi_\beta x, (D_{\varphi_\alpha x}(\varphi_\beta \varphi_\alpha^{-1}))Y)$ whenever $x \in U_\alpha \cap U_\beta$, $Y \in E_\alpha$. Let TM be the space thus obtained, $\widetilde{U}_\alpha$ the canonical image of $(\varphi_\alpha U_\alpha) \times E_\alpha$ in TM, and $\widetilde{\varphi}_\alpha$ the inverse map $\widetilde{U}_\alpha \mapsto (\varphi_\alpha U_\alpha) \times E_\alpha$. The $(\widetilde{U}_\alpha, \widetilde{\varphi}_\alpha, (\varphi_\alpha U_\alpha) \times E_\alpha)$ are charts of a C^{r-1} atlas of TM, which thus becomes a C^{r-1} manifold if $r \geq (1, 1)$, a C^0 manifold otherwise.[12]

If $\widetilde{\varphi}_\alpha X = (\varphi_\alpha x, Y)$, we write $\pi X = x$; this definition does not depend on the choice of α, and $\pi : TM \mapsto M$ is of class C^{r-1} or C^0 (as above). The set $T_x M = \pi^{-1} x$ is identified with E_α by $\widetilde{\varphi}_\alpha$ for each U_α

[10] See Milnor [1].

[11] See, for instance, Palis and de Melo [1] for the finite dimensional version, Abraham and Robbin [1] for generalizations.

[12] If $r = (1, \alpha)$, $0 < \alpha < 1$, we can only say that TM is a C^0 manifold, because $C^{(0,\alpha)}$ manifolds are not defined. See however below the definition of a Hölder bundle over a C^r manifold, $r \geq (0, 1)$.

containing x. The Banach norms obtained on $T_x M$ for different choices of α are equivalent.[13]

One calls *tangent bundle* of M the manifold TM equipped with the map $\pi : TM \mapsto M$, and an atlas of the kind described above. The space $T_x M$ is the *tangent space* to M at x, and for any $\Lambda \subset M$, we write $T_\Lambda M = \pi^{-1}\Lambda$.

Let $\{(V_\beta, \psi_\beta, F_\beta)\}$ be an atlas for a C^r manifold N, and $f : M \mapsto N$ be a C^r map. We define $\widetilde{V}_\beta, \widetilde{\psi}_\beta, TN$ in the obvious manner. There is a map $Tf : TM \mapsto TN$ called *tangent map* with restriction[14]

$$\widetilde{\psi}_\beta^{-1} \circ (\psi_\beta \circ f \circ \varphi_\alpha^{-1}, D(\psi_\beta \circ f \circ \varphi_\alpha^{-1})) \circ \widetilde{\varphi}_\alpha$$

to $U_\alpha \cap f^{-1}V_\beta$ for all α, β. Remark that with this notation $T\varphi_\alpha = \widetilde{\varphi}_\alpha$. The map Tf is C^{r-1} (we shall arrange for this to be true even if $\mathbf{r}-1 = (0, \alpha)$, see below); Tf has a linear restriction $T_x f : T_x M \mapsto T_{fx} N$. If f, g are C^r maps and $g \circ f$ is defined, then $T(g \circ f) = (Tg) \circ (Tf)$. The tangent map to the identity on M is the identity on TM.

If $f : U \mapsto F$ is a function on an open subset of a Banach space E, then $Tf : U \times E \mapsto F \times F$ is defined by

$$(Tf)(x, X) = (fx, (D_x f)X).$$

Thus, Tf plays the role for maps of manifolds, of the derivative Df for maps of Banach spaces.

A C^s *vector field* on M, $\mathbf{s} \leqslant \mathbf{r} - 1$, is a map $X : M \mapsto TM$ such that $\pi X(x) = x$, and if (U, φ, E) is a chart of M, then $(T\varphi) \circ X \circ \varphi^{-1} : \varphi(U) \mapsto \varphi(U) \times E$ is of class C^s.

One can give a general definition of a C^s *vector bundle* $\widetilde{M}$ over the C^r manifold M, where $\mathbf{r} = 0$ or $\mathbf{r} \geqslant (0, 1)$ and $\mathbf{s} \leqslant \mathbf{r}$. Essentially, $\widetilde{M}$ is a set with a map $\pi : \widetilde{M} \mapsto M$, such that for an open cover $\{U_\alpha\}$ of M and Banach spaces E_α, a bijection $\pi^{-1}U_\alpha \mapsto U_\alpha \times E_\alpha$ is given, with π corresponding to the projection on the first factor. The

---

[13]$\pi^{-1}x$ is thus a "Banachable space." Many mathematicians prefer to say "Banach space," keeping in mind the fact that the norm is defined only up to equivalence.

[14]It is understood in this formula that

$$(\psi_\beta \circ f \circ \varphi_\alpha^{-1}, D(\psi_\beta \circ f \circ \varphi_\alpha^{-1}))(\varphi_\alpha x, Y) = (\psi_\beta f x, D_{\varphi_\alpha x}(\psi_\beta \circ f \circ \varphi_\alpha^{-1})Y).$$

maps $U_\alpha \cap U_\beta \mapsto \mathcal{L}(E_\alpha, E_\beta)$ thus defined are assumed to be C^s; in this manner $\widetilde{M}$ becomes a C^s manifold if $s \geqslant (0,1)$, a C^0 manifold otherwise.[15] The linear spaces $\pi^{-1}x$ are called *fibers*.

If M is a topological space, one can still define *continuous vector bundles* on M; if M has a metric, one can define *Hölder vector bundles* ($C^{(0,\alpha)}$ bundles with $\alpha \in [0,1]$).

Suppose now that $(\widetilde{N}, \varpi)$ is a C^s vector bundle over N. A map $\tilde{f}$: $\widetilde{M} \mapsto \widetilde{N}$ is a C^s *vector bundle map* over $f : M \mapsto N$ if $\varpi \circ \tilde{f} = f \circ \pi$, and $\tilde{f}$ corresponds, via bijections $x^{-1}U_\alpha \mapsto U_\alpha \times E_\alpha$, $\varpi^{-1}V_\beta \mapsto V_\beta \times F_\beta$, to C^s functions $U_\alpha \cap f^{-1}V_\beta \mapsto \mathcal{L}(E_\alpha, F_\beta)$. We denote by $\tilde{f}_x : \pi^{-1}x \mapsto \varpi^{-1}fx$ the linear map induced by $\tilde{f}$ on fibers.

Let $\tilde{f} : \widetilde{M'} \mapsto \widetilde{M}$ be a C^s vector bundle map over the identity map of M, such that $\tilde{f}_x$ is the inclusion map of a closed subspace $\pi'^{-1}x$ with closed complement in $\pi^{-1}x$. Then we call $\widetilde{M'}$ a C^s *subbundle* of $\widetilde{M}$.

The *product* $\widetilde{M} \times \widetilde{N}$ of vector bundles $\widetilde{M}$ over M and $\widetilde{N}$ over N is a vector bundle over $M \times N$ defined by bijections $\pi^{-1}U_\alpha \times \varpi^{-1}V_\beta \mapsto (U \times V) \times (E_\alpha \times V_\beta)$ in the obvious manner.

A C^s *section* of the bundle $\widetilde{M}$ is a map $X : M \to \widetilde{M}$ such that $\pi X(x) = x$, and the bijections $\pi^{-1}U_\alpha \mapsto U_\alpha \times E_\alpha$ transform X into C^s functions $U_\alpha \mapsto E_\alpha$.

In the above terminology, the tangent bundle to a C^r manifold, $r \geqslant 1$, is a C^{r-1} vector bundle. Given a C^r map $f : M \mapsto N$, Tf is a C^{r-1} vector bundle map over f (even if $r = (1,\alpha)$). The tangent bundle $T(M \times N)$ is the product bundle $TM \times TN$. A C^s vector field on M is the same thing as a C^s section of TM.

If H is a subset of M, we write $T_H M = \pi^{-1}H$. If H is a submanifold, $T_H M$ is thus a vector bundle over H.

B.6. *Differential Equations*

Let M be a C^r manifold, $r \geqslant 1$. Given an open interval $J \subset \mathbf{R}$ and a smooth function $f : J \mapsto M$, we have a map $Tf : TJ \mapsto TM$. With

---

[15]More precisely, the bijections $\pi^{-1}U_\alpha \mapsto U_\alpha \times E_\alpha$ define a "vector atlas" on $\widetilde{M}$. Two vector atlases are equivalent if their union is again a vector atlas, and $\widetilde{M}$ equipped with an equivalence class of vector atlases is a C^s vector bundle. Hölder vector bundles of exponent α can be defined (whereas Hölder manifolds of exponent α cannot). This is because the equivalence of vector atlases involves considering only products of (operator-valued) Hölder functions (rather than composition of Hölder functions).

the identification $TJ = J \times \mathbf{R}$, we write

$$f'(t) = (Tf)(t, 1).$$

If X is a C^s vector field on M, $s \leq r - 1$, one says that f is an *integral curve* of X, or a *solution* of the differential equation

(B.3) $$f' = X(f)$$

in J, provided $f'(t) = X(f(t))$ for all $t \in J$; f is then of class C^{s+1}. One often writes $\frac{df}{dt}$ for $f'(t)$ or (improperly) for f'. The differential equation (B.3) then becomes

$$\frac{df}{dt} = X(f).$$

More generally, one may consider the time-dependent equation

(B.4) $$\frac{df}{dt} = \widetilde{X}(f, t)$$

with a time-dependent vector field $t \mapsto X(\cdot, t)$ defined for $t \in \widetilde{J}$. Introduce a new vector field $\widetilde{X} : (x, t) \mapsto (X(x, t), (t, 1))$ on $M \times J$. Then f is a solution of (B.4) if and only if $\widehat{f} : t \mapsto (f(t), t)$ is a solution of the time-independent equation $\frac{d\widehat{f}}{dt} = \widetilde{X}(\widehat{f})$.

We return now to the study of time-independent equations.

B.6.1. Theorem (on the flow defined by a vector field).[16] *Let X be a C^s vector field on the C^r manifold M, and suppose that $(0, 1) \leq s \leq r - 1$. Given $x \in M$, there is some interval $J \ni 0$ and some solution f of (B.3) in J such that $f(0) = x$. Furthermore, any two solutions equal at 0 are equal on the intersection of their intervals of definition. There is thus a largest interval $(T_-(x), T_+(x))$ where a solution f with $f(0) = x$ is defined. We denote this solution by $t \mapsto f^t x$. The set $\Gamma = \{(x, t) \in M \times \mathbf{R} : T_-(x) \leq t \leq T_+(x)\}$ is open in $M \times \mathbf{R}$, and the map $(x, t) \mapsto f^t x$ is C^s on Γ. The property $f^s \circ f^t = f^{s+t}$ holds where it makes sense. If M is compact, then $\Gamma = M \times \mathbf{R}$.*

If the manifold M is just an open subset U of a Banach space E, a vector field on M can be identified with a function $U \mapsto E$. By keeping

---

[16] The $C^{(0,1)}$ case, the C^r case for $r \geq 1$ integer, and the C^∞ case are proved in Lang [1]. The method used by Lang in the C^r case r integer, extends easily to the $C^{(r,\alpha)}$ case. The results for the C^ω case are in Bourbaki [5]; this case can be treated by use of power series.

the above notation and assuming s ⩾ 1, one finds that the function $t \mapsto D_x f^t$ is a solution of the linear equation

(B.5) $$\frac{d}{dt}(D_x f^t) = (D_{f^t x} X).(D_x f^t)$$

in $\mathcal{L}(E, E)$. This equation is time dependent and contains x as a parameter.

B.7. Riemann Metrics

Let H be a Hilbert space, and M a C^{r+1} manifold of type H (see Section B.4: M is a C^{r+1} Hilbert manifold). Suppose that for each $x \in M$, a Hilbert scalar product $(\ ,\ )_x$ is defined on $T_x M$. Suppose also that if (U, φ, H) is a C^{r+1} chart, there is a C^r function $x \mapsto A_x \in \mathcal{L}(E)$ such that

$$(u, v)_x = ((T_x \varphi)u, A_x (T_x \varphi)v).$$

Then, the collection of scalar products $(\ ,\ )_x$ defines a C^r *Riemann metric* on M. (For each x, the operator A_x is positive and invertible.)

If M is a separable C^{r+1} Hilbert manifold, $r \leqslant \infty$, it has a C^r Riemann metric. (This follows from the existence of partitions of unity, Proposition B.4.2). For instance, a compact C^∞ manifold has a C^∞ Riemann metric.

If $\alpha : [0, 1] \mapsto F$ is smooth, the length of the curve α is defined by

$$\ell(\alpha) = \int_0^1 [(\alpha'(t), \alpha'(t))_{\alpha(t)}]^{1/2} dt,$$

and the distance between $x, y, \in M$ is

$$d(x, y) = \inf_\alpha \{\ell(\alpha) : \alpha(0) = x, \alpha(1) = y\}.$$

[This definition of length and distance extends to the case of a C^1 Banach manifold M with norm $\| \cdot \|_x$ on T_x depending continuously on x.]

Given a C^{r+1} function $\Phi : M \mapsto \mathbf{R}$, a C^r vector field X is defined by

$$(T_x \Phi)(u) = (X(x), u)_x.$$

X is called the *gradient* of Φ with respect to the Riemann metric. The flow corresponding to the gradient vector field is the *gradient flow*. Note that for a gradient flow, the function $t \mapsto \Phi(f^t x)$ is always nondecreasing.

B.8. Note

The material in this appendix is mostly extracted from Lang [1] and
Bourbaki [5], which should be consulted for more information. We
have also systematically introduced the $C^{(r,\alpha)}$ differentiability classes,
because they naturally appear in the study of differentiable dynamical
systems. For a general definition of differentiability classes, see Bonic
and Frampton [1].

The Leray–Schauder degree introduced in Section B.2 is important
in the topological study of dynamical systems. We insist however more
on differentiable techniques. The main nontrivial results here are the
implicit function theorem (B.3.1 or B.3.3) and Theorem B.6.1 on differential equations. The implicit function theorem can be proved by a
contraction mapping argument. The theorem on the flow defined by a
differential equation can be obtained in various ways, including application of the implicit function theorem.

C. Topological Dynamics and Ergodic Theory

The basic object of topological dynamics is a topological space M with
a group or semigroup of continuous maps $M \mapsto M$. The basic object of
ergodic theory is a probability space (M, ρ) with a group or semigroup
of measure-preserving maps. These structures underlie differentiable
dynamics, whence their interest here.

C.1. Topological Transitivity and Mixing

Let M be a nonempty Hausdorff topological space, and (f^t) a group or
semigroup of maps $f^t : M \mapsto M$. The "time" t takes values in $\mathbf{Z}$ or $\mathbf{R}$,
possibly restricted by $t \geq 0$, and we have $f^{s+t} = f^s \circ f^t$, $f^0 = $ identity.
We assume that $(x, t) \mapsto f^t x$ is continuous.

A point $x \in M$ is *wandering* if there are an open set $U \ni x$ and $T > 0$
such that $f^t U \cap U = \emptyset$ for $t \geq T$; otherwise x is *nonwandering*. The
nonwandering set Ω (i.e., the set of nonwandering points) is closed, and
$f^t \Omega \subset \Omega$.

We say that (f^t) is *topologically + transitive* or *topologically mixing*
respectively when the following condition $(+T)$ or (M) is satisfied.

$(+T)$ *If U, V are nonempty open sets and $T > 0$, there exists $t \geq T$
        such that $f^t U \cap V \neq \emptyset$.*

(M) *If U, V are nonempty open sets, there exists $T > 0$ such that*

$f^t U \cap V \neq \emptyset$ *for all* $T \geqslant T$.

Condition $(+T)$ implies that $M = \Omega$; condition (M) implies $(+T)$.

If f^t is defined for $t < 0$, time reversal (i.e., $t \mapsto -t$) does not change the nonwandering set Ω nor the conditions $(+T)$, (M). If f is defined for $t < 0$, one says that (f^t) is *topologically transitive* if the following condition (T) is satisfied.

(T) *If U, V are nonempty sets, there exists t (in $\mathbf{Z}$ or $\mathbf{R}$) such that* $f^t U \cap V \neq \emptyset$.

Condition $(+T)$ is equivalent to $((T)$ and $M = \Omega)$.

If M is compact, the preorder $\succ$ and the *chain-recurrent set* are defined. (See Section 8.3, and note that there is a unique uniform structure compatible with the topology of M.) The chain-recurrent set R is closed, and $\emptyset \neq \Omega \subset R$.

If M is compact metrizable, then $(+T)$ is equivalent to the following conditions.

$(+T')$ *There is $x \in M$ such that $\{f^t x : t \geqslant 1\}$ is dense in M.*

$(+T'')$ *There is a residual subset of M consisting of points x such that $\{f^t x : t \geqslant 1\}$ is dense in M.*

If M is compact metrizable and f^t is defined for $t < 0$, then (T) is equivalent to the following conditions.

(T') *There is $x \in M$ such that $\{f^t x\}$ is dense in M.*

(T'') *There is a residual subset of M consisting of points x such that $\{f^t x\}$ is dense in M.*

C.2. Ergodic Theory

There are different approaches to measure theory; see, in particular, Halmos [1] and Bourbaki [4], to which we refer for the definitions of measures, measurability, and integrability. If ρ is a measure, we denote by $\rho(E)$ the measure of a set E, and by

$$\rho(\varphi) = \int \varphi(x)\rho(dx)$$

the integral of a function φ.

We recall that a measure ρ on M is a *probability measure* if it is positive $(\rho \geqslant 0)$ and normalized (i.e., $\rho(M) = 1$, or $\rho(1) = 1$, where 1 denotes the constant function with value 1 on M). If (f^t) is a family of measurable maps $M \mapsto M$, we say that the measure ρ is *invariant*—or (f^t) *invariant*—if $\rho(A) = \rho((f^t)^{-1}A)$ for all t and all measurable sets

A. (In particular, for a discrete dynamical system generated by the map f, we have f-invariance, i.e., $\rho(A) = \rho(f^{-1}A)$ for all measurable A.) An invariant probability measure ρ is *ergodic* if there is no nontrivial decomposition

$$\rho = \alpha\rho_1 + (1 - \alpha)\rho_2$$

(i.e., no such representation of ρ where $\alpha \neq 0$ or 1, and ρ_1, ρ_2 are distinct invariant probability measures).

C.2.1. Theorem (Ergodic theorem of Birkhoff). *Let ρ be a probability measure on M, and $f : M \mapsto M$ a measurable map such that ρ is f-invariant. If $\varphi \in L^1(M, \rho)$, (i.e., if φ is integrable), there is a set $E \subset M$ of full measure (i.e., $\rho(E) = 1$) such that the following limit exists for all $x \in E$,*

$$\lim_{n \to \infty} \frac{1}{n} \sum_{k=0}^{n-1} \varphi(f^k x) = \overline{\varphi}(x).$$

If ρ is ergodic, $\overline{\varphi}$ is ρ-almost-everywhere constant, and equal to $\rho(\varphi)$.

Suppose now that M is compact and metrizable.[17] The measures that we shall consider are the so-called *Borel measures* (see the above references); we simply call them *measures* in what follows. In this situation, a probability measure ρ on M defines a linear functional

$$\varphi \to \rho(\varphi) = \int \varphi(x)\rho(dx)$$

on the space of continuous functions $\varphi : M \mapsto \mathbf{C}$. This linear functional is positive ($\rho(\varphi) \geqslant 0$ if $\varphi \geqslant 0$) and normalized ($\rho(1) = 1$). Conversely, every positive normalized linear functional corresponds in this manner to a probability measure. The *support* of ρ (denoted by $\operatorname{supp}\rho$) is the smallest closed set K such that $\rho(\varphi) = 0$ if φ vanishes on K. The f-invariance of ρ can be formulated as $\rho(\varphi) = \rho(\varphi \circ f)$ for all continuous $\varphi : M \mapsto \mathbf{C}$. The ergodic theorem now takes the following form.

C.2.2. Corollary (Ergodic theorem on a compact metrizable space). *Let f be a continuous map of the compact metrizable space M to itself. There*

---

[17]For differentiable dynamics, this is the situation of interest, because one usually considers measures with compact support in a manifold that is metrizable.

is an invariant set E such that $\rho(E) = 1$ for every invariant probability measure ρ, and the following limit exists,

$$\lim_{n\to\infty} \frac{1}{n} \sum_{k=0}^{n-1} \varphi(f^k x) = \overline{\varphi}(x)$$

for all continuous functions $\varphi : M \mapsto \mathbf{C}$ and all $x \in E$.

[This follows from C.2.1 because the space of continuous functions $M \mapsto \mathbf{C}$ with the topology of uniform convergence contains a countable dense set.]

If $x \in E$, the map $x \mapsto \overline{\varphi}(x)$ defines a probability measure ρ_x such that $\rho_x(\varphi) = \overline{\varphi}(x)$; ρ_x is f-invariant. The set E in C.2.2 may be chosen such that *all the measures ρ_x are ergodic.* This results from the Bogoliubov–Krylov theory, which in effect gives a decomposition of ρ into its ergodic components ρ_x (see Jacobs [1] for a discussion of the Bogoliubov–Krylov theory). Another way of looking at the ergodic decomposition of ρ is provided by the Choquet theory (see Choquet et Meyer [1]). In particular, the support of an invariant probability measure ρ is contained in the closure of the union of the supports of the ergodic probability measures ρ_x. If $y \in \operatorname{supp} \rho_x$, then $f^k x$ comes arbitrarily close to y for arbitrarily large k, i.e., y is nonwandering. In conclusion, the support of all ergodic measures (and therefore all invariant measures) is contained in the nonwandering set.

C.3. Note

The following are useful monographs on topological dynamics and ergodic theory: Denker, Grillenberger, and Sigmund [1], and Walters [1].

D. Axiom A Dynamical Systems

This appendix discusses an important special class of dynamical systems, the Axiom A systems, for which the geometry of complex dynamical behavior can be analyzed in detail.

D.1. Definitions

Let f be a $C^{\mathbf{r}}$ map or (f^t) a $C^{\mathbf{r}}$ semiflow of the $C^{\mathbf{r}}$ manifold M, where $\mathbf{r} \geqslant 1$ ($\mathbf{r} > 1$ if $\dim M = \infty$).[18] We assume that $(x, t) \mapsto f^t x$ is continuous for $t \geqslant 0$ and $C^{\mathbf{r}}$ for $t > 0$. Let Ω be the nonwandering set; by definition,[19] $x \notin \Omega$ means that there are a neighborhood $\mathcal{N}$ of x and $T > 0$ such that $\mathcal{N} \cap f^t \mathcal{N} = \emptyset$ when $t > T$. The nonwandering set Ω is closed.

We say that f is an *Axiom A map* or $f(t)$ an *Axiom A semiflow* if the nonwandering set Ω is compact and the following conditions are satisfied:

(Aa) *Ω is hyperbolic* (see Section 2.7),
(Ab) *the fixed points and periodic orbits are dense in Ω*.

These definitions are due to Smale [3] in the case of a diffeomorphism or flow on a compact manifold M. Examples are known of diffeomorphisms of a compact manifold satisfying (Aa) but not (Ab).[20]

As usual, we may cover the compact set Ω by a finite number of $C^{\mathbf{r}}$ charts. For x in a neighborhood U of Ω, we then define a norm on $T_x M$, depending continuously on x and equivalent to the norms induced by the charts. We also define a distance d on U such that it differs by a bounded factor from the norm distances induced by the charts.[21]

Since Ω is hyperbolic, stable and unstable manifolds are defined for $x \in \Omega$ (Section 15.2), and the dynamical system restricted to Ω is expansive (Section 15.3).

D.2. Theorem (Local product structure).[22] *The nonwandering set Ω for an Axiom A dynamical system has local product structure.*

We shall prove this in the case of a map. Remember that, if $x \in \Omega$,

$$V_x^- = \{y \in M : d(f^n x, f^n y) < R \quad \text{for } n \geqslant 0\},$$
$$V_x^+ = \{y_0 \in M : \exists (y_k)_{k \leqslant 0}, \quad d(f^{-n} x, y_{-n}) < R,$$
$$\text{and } f y_{-n-1} = y_{-n} \quad \text{for} \quad n \geqslant 0\}.$$

---

[18] We need $\mathbf{r} > 1$ if $\dim M = \infty$ because of our use of the λ-lemma in the proof of Theorem D.2.

[19] See Appendix C.1.

[20] See Dankner [1], Kurata [1].

[21] The way to do this is explained in Appendix B.7.

[22] See Smale [3].

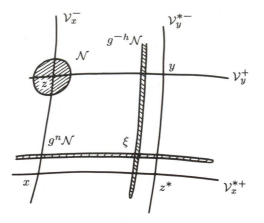

FIG. 38.   Cloud lemma: if x and y are hyperbolic periodic points and $z \in \mathcal{V}_x^- \cap \mathcal{V}_y^+$, then z is nonwandering. This is because the images $g^n \mathcal{N}, g^{-n} \mathcal{N}$ of the *cloud* $\mathcal{N}$ around z have a point in common near z^*.

Similarly, we define $\mathcal{V}_x^{*\pm}$ with R^* replacing R. We assume R, R^* sufficiently small for the $\mathcal{V}, \mathcal{V}^*$ to be "nearly flat." Furthermore, given R^*, we take R small enough that

$$(\mathcal{V}_y^{*-} \cap \mathcal{V}_x^{*+} = \emptyset) \Rightarrow (\mathcal{V}_x^- \cap \mathcal{V}_y^+ = \emptyset).$$

Equivalently, if $z \in \mathcal{V}_x^- \cap \mathcal{V}_y^+$, there is a point of intersection $z^* \in \mathcal{V}_y^{*-} \cap \mathcal{V}_x^{*+}$, see Fig. 38.

What we have to prove (see Section 15.4) is that $z \in \Omega$. Since the periodic points are dense in Ω, since $\mathcal{V}_x^-$ and $\mathcal{V}_y^+$ depend continuously on x, y, and since Ω is closed, it suffices to show that $z \in \Omega$ when x and y are periodic. By taking $g = f^N$ for suitable N, we may assume that $gx = x$, $gy = y$, and we want to prove that for every neighborhood $\mathcal{N}$ of z, there are arbitrarily large n such that $g^n \mathcal{N}$ and $g^{-n} \mathcal{N}$ have nonempty intersection. Intuitively, this is clear (see Fig. 38): $g^n \mathcal{N}$ contains a thin "pancake" near $\mathcal{V}_x^{*+}$, and $g^{-n} \mathcal{N}$ contains a thin "pancake" near $\mathcal{V}_y^{*-}$, therefore $g^n \mathcal{N} \cap g^{-n} \mathcal{N}$ contain a point ζ near z^*.

A rigorous version of the above argument is known as the *cloud lemma*.[23] This is an easy consequence of the inclination lemma (Problem 5 of Part 1). In fact, $g^n (\mathcal{N} \cap \mathcal{V}_y^+)$, suitably restricted, tends to $\mathcal{V}_x^{*+}$, while $g^{-n} (\mathcal{N} \cap \mathcal{V}_x^-)$ tends to $\mathcal{V}_y^{*-}$. Both limits are in the C^1 topology

[23] See Smale [3].

(see Problem 5, Part 1), and therefore $g^n \mathcal{N}$ and $g^{-n} \mathcal{N}$ have a point of intersection near z^*.

The proof for semiflows is similar.

D.3. Theorem (Spectral decomposition). *The nonwandering set Ω of an Axiom A map f or semiflow (f^t) is a finite union*

(D.1) $$\Omega = \Omega_1 \cup \cdots \cup \Omega_N$$

of pairwise disjoint closed invariant sets, such that f or (f^t) is topologically transitive on each Ω_i.

The Ω_i are called *basic sets*. Remember that topological transitivity means that there is an x with dense orbit in Ω_i.[24] In fact, there is a residual set of such x, and this implies that *the spectral decomposition* (D.1) *is unique*. It is also clear that *in the semiflow case, the basic sets are connected*. The spectral decomposition is due to Smale; the proof is not hard.[25] [If U is a small open set intersecting Ω, the closure of $\bigcup_{t \geqslant 0} f^t(U \cap \Omega)$ is a basic set Ω_i. The essential thing is to prove (by using local product structure and the density of periodic points) that, for smaller $V \subset U$, the set $\bigcup_{t \geqslant 0} f^t(V \cap \Omega)$ is dense in U, and therefore its closure is again Ω_i.]

If x is a periodic point of period T, let us write

(D.2) $$\Omega_x^* = \text{closure} \bigcup_{n \geqslant 0} f^{nT} V_x^+ \cap \Omega),$$

where the union is over the positive integers n. (Ω_x^* is thus the closure of the global unstable manifold of x restricted to Ω.)

D.4. Theorem (Decomposition of basic sets for maps). *Each basic set Ω_i for an Axiom A map is a finite union $\Omega_i = \Omega_{i1} \cup \cdots \cup \Omega_{iN_i}$ of pairwise disjoint closed sets of the form* (D.2) *which are permuted cyclically by f, and such that f^{N_i} is topologically mixing on Ω_{ij}. This decomposition is unique.*

Topological mixing is defined in Appendix C.1. [By using the local product structure one shows that if y is periodic and close enough to Ω_x^*, then $y \in \Omega_x^*$, hence Ω_x^* is open and closed in Ω_i, and the rest follows easily, see Bowen [2].]

---

[24] See Appendix C.1. Since Ω_i is nonwandering, topological transitivity and + transitivity coincide. One may thus use equivalently full orbits or future orbits in the definition.

[25] See Smale [3], and Pugh and Shub [2] for the flow case.

D.5. Theorem (Decomposition of basic sets for semiflows). *Let Ω_i be a basic set for an Axiom A semiflow, with $x \in \Omega_i$ and Ω_x^* defined by* (D.2). *One of the following three mutually exclusive possibilities holds:*

(a) *x is a fixed point, and $\Omega_i = \{x\}$.*

(b) *$\Omega_x^* \neq \Omega_i$, and there is $T > 0$ such that the flow $(f^t|\Omega_i)$ is the suspension of $f^T|\Omega_x^*$ with constant roof function T.*

(c) *$\Omega_x^* = \Omega_i \neq \{x\}$; W_x^+ is thus dense in Ω_i, and the flow $(f^t|\Omega_i)$ is topologically mixing.*

In case (b), the restriction of f^T to Ω_x^* is a priori only a homeomorphism; the suspension with constant roof function T is defined as in Section 1.4. [For the proof of this theorem see Bowen [3].]

D.6. Local Definition of Axiom A Basic Sets, Attractors

A local, and possibly more general, definition of a basic set may be given. Let Λ be a compact subset of M, U an open neighborhood of Λ. We assume that $f : U \mapsto M$ is a $C^{\mathbf{r}}$ map. Or we assume that $(x, t) \mapsto f^t x$ is continuous: $U \times [0, T) \mapsto M$ and $C^{\mathbf{r}}$: $U \times (0, T) \mapsto M$, such that $f^0 = $ identity and $f^{s+t} = f^s \circ f^t$ where defined; the definition of f^t is then naturally extended to all $t \geqslant 0$, and (f^t) is a local semiflow.

We say that Λ is an *Axiom A basic set* if the following conditions are satisfied.

(a) $f^t \Lambda = \Lambda$ *for all t* (invariance).

(b) Λ *is a hyperbolic set* (the definition of Section 2.7 makes sense in the present local setting, in particular, $f^t|\Lambda$ is a homeomorphism).

(c) *Periodic points are dense in Λ.*

(d) *The dynamical system (for (f^t)) is topologically transitive on Λ.*

(e) *If $x_s \in U$ for all $s \in \mathbf{Z}$ or $\mathbf{R}$ and $f^t x_s = x_{s+t}$, then $x_0 \in \Lambda$.*

If Ω_i is a basic set for a globally defined Axiom A system, (a)–(d) are clearly satisfied, and (e) results from the existence of fundamental neighborhoods for sets with local product structure (Sections 15.5, 15.7). The locally defined Axiom A basic sets again have local product structure and satisfy the Decomposition Theorem D.4 or D.5. By topological transitivity, all points of Λ are equivalent in the sense of Section 8.4. Notice that the *unstable dimension* $\dim V_x^+ = \dim \mathcal{V}_x^+$ is independent of $x \in \Lambda$. (The same is true of the *stable dimension* $\dim V_x^-$ for a diffeomorphism or a flow).

A hyperbolic periodic orbit, or the *horseshoe* of Section 16, are examples of Axiom A basic sets.

We say that the Axiom A basic set Λ is an *Axiom A attractor* if U may be chosen such that

(e*)
$$\bigcap_{t \geqslant 0} f^t U = \Lambda.$$

Λ is then an attracting set (Section 8.1) and an attractor (Section 8.4). If the unstable dimension of a basic set is zero, then Λ is an attracting periodic orbit [by (c), Λ contains a periodic orbit, $\dim V^+ = 0$ implies that it is attracting, and (d) that Λ is equal to this attracting periodic orbit]. If an Axiom A attractor has $\dim V^+ > 0$, we have *sensitive dependence on initial condition* (see Section 8.4), and Λ is a *strange Axiom A attractor*. We shall describe examples of such attractors below.

D.7. Proposition (Characterization of attractors). *An Axiom A basic set Λ is an attractor if and only if $V_x^+ \subset \Lambda$ for some $x \in \Lambda$; the unstable manifolds are then all contained in Λ. In particular, $\dim V_x^+ < +\infty$.*

If (e*) holds, we have for small $\varepsilon > 0$,

$$V_x^+ \cap B_x(\varepsilon) \subset \bigcap_{t \geqslant 0} f^t V_{f^{-t}x}^+ \subset \bigcap_{t \geqslant 0} f^t U = \Lambda.$$

By applying f^T with $T > 0$, we find $V_x^+ \subset \Lambda$. Suppose now that (e*) does not hold, for a small neighborhood U of Λ, and let

$$\bigcap_{t \geqslant 0} f^t U \ni y \notin \Lambda.$$

There are pieces of orbit $\{y_s : -T \leqslant s \leqslant 0\}$, with $y_0 = y$ and T arbitrarily large, contained in U. By shadowing (Section 15.5 or 15.7) and the construction of V^+, V^{0+} (Section 15.2), we may assume that y is very close to V_x^+ (or V_x^{0+} for a semiflow) for some $x \in \Lambda$. If y is closer to V_x^+ or V_x^{0+} than to Λ, we obtain $V_x^+ \not\subset \Lambda$. This argument shows that $V_x^+ \not\subset \Lambda$ for $x \in$ nonempty open set S, hence for $x \in \bigcup_{n \geqslant 0} f^{-n} S$, hence for all $x \in \Lambda$ by using the local product structure of Λ.

We have thus shown that

$$(V_x^+ \subset \Lambda \text{ for some } x) \Rightarrow (\Lambda \text{ is an attractor}) \Rightarrow (V_x^+ \subset \Lambda \text{ for all } x).$$

Finally, compactness of Λ implies that the manifold V_x^+ has finite dimension when $V_x^+ \subset \Lambda$.

D.8. Stable Manifolds of an Attractor

The stable and unstable manifolds $\mathcal{V}_x^{\pm}$ of an Axiom A basic set Λ depend continuously on x, and we also know that $x \mapsto \mathcal{V}_x^{\pm}$ is Hölder continuous under suitable conditions (Problem 4 of Part 2). Can we say more in the case of an attractor?

If Λ is an attractor and $x \in \Lambda$ then $\mathcal{V}_x^+ \subset \Lambda$, and in the case of a map, the $\mathcal{V}_y^-$ with $y \in \mathcal{V}_x^+$ cover a neighborhood of x [use the existence of a fundamental neighborhood of Λ]. In the case of a semiflow, the $\mathcal{V}_y^-$ with $y \in \mathcal{V}_x^{0+}$ cover a neighborhood of x. *The map $y \mapsto \mathcal{V}_y^-$ is usually not smooth.* In fact, according to the general philosophy of normal hyperbolicity (see Section 14), we expect $y \mapsto V_x^-$ to be C^1 if a subbundle of $T_\Lambda M$, close to V^-, will approach V^- under iterates of Tf^{-1} faster than distances on the unstable manifolds are contracted by iterates of f^{-1}. Using this argument the following may be proved:[26]

Let Λ be an Axiom A attractor with unstable dimension $\dim V^+ = 1$ for the C^2 map f. If $x \in \Lambda$, the map $\mathcal{V}_x^+ \ni y \mapsto V_y^-$ is C^1. In fact, the V_y^- form a C^1 foliation of a neighborhood $\mathcal{O}$ of x, i.e., there is a C^1 chart $\varphi : \mathcal{O} \mapsto \mathbf{R} \times F$ such that $\varphi V_y^- \subset \{\varphi_1 y\} \times F$.

D.9. Pre-Axiom A Dynamical Systems: Expanding Maps

Let K be a prehyperbolic set with dense periodic points. Then, the periodic points are also dense in the hyperbolic cover $K^\dagger$. [If $(x_s)_{s\leqslant0} \in K^\dagger$, choose y periodic such that $d(f^t y, f^t x_{-T}) < \varepsilon$ for large T and small ε.]

This suggests defining *pre-Axiom A dynamical systems* such that the nonwandering set Ω is compact and satisfies

(Aa)† Ω *is prehyperbolic*

(Ab)   *the fixed points and periodic orbits are dense in Ω.*

Similarly, we obtain a local definition of *pre-Axiom A basic sets* if in Section D.6, we replace (b) by

(b)† Λ *is a prehyperbolic set.*

The proof of Theorem D.2 applies to pre-Axiom A dynamical systems, so that $\Omega^\dagger$ (resp. $\Lambda^\dagger$) has local product structure. The spectral decomposition and the decomposition of basic sets therefore extend to the present situation. The persistence properties of sets with local

---

[26] See Hirsch and Pugh [1], Theorem 6.5, where $\dim V^+ = 1$ is replaced by a weaker condition. See also the C^1 section theorem (6.5) in Hirsch, Pugh, and Shub [1].

product structure apply here also (see Sections 15.5, 15.6, 15.7).

An interesting class of pre-Axiom A systems is provided by expanding maps. We say that a smooth map $f : M \mapsto M$ of a compact manifold M is an *expanding map* if $fM = M$, and there are $\theta > 1$ and a Riemann metric such that[27]

(D.3) $$\|Tfu\| \geqslant \theta\|u\|$$

for all $u \in TM$. Clearly, M is then prehyperbolic, and the stable manifolds are reduced to points. One can show that periodic points are dense in M, and therefore f is a pre-Axiom A map. The following *persistence property* holds: if $\tilde{f}$ is C^1 close to f, then $\tilde{f}$ is again expanding, and there is a homeomorphism $h : M \mapsto M$ such that $\tilde{f} \circ h = h \circ f$.

The simplest examples of expanding maps are the maps $\theta \mapsto N\theta \pmod 1$ of the circle S^1 for integer $N > 1$.[28]

D.10. An Example of Attractor: the Solenoid

Let S^1 be the circle and $g : \theta \mapsto 2\theta \pmod 1$ the *circle doubling* map. Suppose that S^1 is identified with a submanifold S of a compact manifold M of dimension 3. For suitable M, S, we may extend g to a smooth map $f : M \mapsto M$ that normally contracts to S. Specifically, we may assume that there is a Riemann metric on M, and that the *normal bundle*

$$\{u \in T_S M : u \text{ is orthogonal to } S\}$$

is sent to itself and each vector contracted by Tf. In view of what follows, we assume that this normal contraction is fairly strong.

Clearly, the set S is prehyperbolic and, in fact, a pre-Axiom A basic set. Its hyperbolic cover $S^\dagger$ consists of sequences $(\theta_k)_{k \leqslant 0}$ such that $g\theta_{k-1} = \theta_k$, i.e., $2\theta_{k-1} = \theta_k \pmod 1$. Let I_0, I_1 be the intervals $[0, \frac{1}{2}], [\frac{1}{2}, 1]$ of S. There is an almost one-to-one correspondence between points of $S^\dagger$ and sequences $(\xi_k)_{k \in \mathbf{Z}}$ established as follows:

$$\xi_k = i \quad \text{if} \quad \begin{cases} g^k \theta_0 \in I_i & \text{for } k \geqslant 0 \\ \theta_k \in I_i & \text{for } k \leqslant 0 \end{cases}$$

(there is an ambiguity when $g^k\theta_0$ or θ_k is 0 or $\frac{1}{2}$). The action of $g^\dagger$ on $S^\dagger$ corresponds to the *shift* of (ξ_k) by one place to the left. As in

---

[27]A weaker condition is sufficient: if there are $C > 0$, $\theta > 1$ such that $\|Tf^n u\| \geqslant C\theta^n \|u\|$ for all $n \geqslant 0$ and $u \in TM$, one can find a new metric such that (D.3) holds.

[28]For more details on expanding maps, see Shub [1], Hirsch [1].

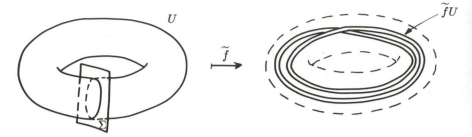

FIG. 39.   The solenoid. The map $\tilde{f}$ has an Axiom A attractor called a solenoid.

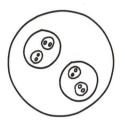

FIG. 40.   The Cantor set structure of a cross-section through a solenoid.

Section 16.1, we see that the original differentiable dynamics is equivalent to a *symbolic dynamics* defined by the shift on sequences of symbols.

By a small perturbation of f, we may produce a map $\tilde{f}$ that sends a neighborhood U of S^1 injectively to itself as shown by Fig. 39.[29] The neighborhood U is a solid torus, and $\tilde{f}U$ winds twice around the central hole of U. (We need a strong normal contraction so that $\tilde{f}U$ fits inside U.) In general, $\tilde{f}^n U$ winds 2^n times around the central hole of U, and the intersection $\Lambda = \bigcap_{n \geqslant 0} \tilde{f}^n U$ is a wiry structure called a *solenoid*. By the persistence of pre-Axiom A sets, Λ is an Axiom A basic set ($\tilde{f}|\Lambda$ is injective); it is in fact an attractor. Since $\dim V^+ = 1$, the solenoid is a *strange Axiom A attractor*.

If we cut U by a piece of surface Σ, then $\Sigma \cap \tilde{f}^n U$ will appear as a union of 2^n disjoint disks (see Fig. 40). Therefore, the points of $\Sigma \cap \Lambda$ correspond to a sequence of binary choices, and $\Sigma = \cap \Lambda$ is a Cantor set. The occurrence of Cantor sets is a feature of many strange attractors (or, in general, basic sets) and accounts for some of their strangeness.[30]

---

[29] For suitable f, we may assume that $\tilde{f}$ is a diffeomorphism of M.

[30] Note that Guckenheimer [1] has given an example of an Axiom A basic set that is not locally the product of a manifold and a Cantor set.

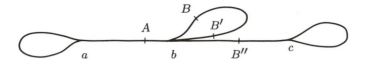

FIG. 41. A compact connected 1-dimensional branched manifold S. It is obtained by gluing five 1-cells at the three branch points a, b, c. Smoothness around b may be defined using the 1-cells AB, AB', and AB''.

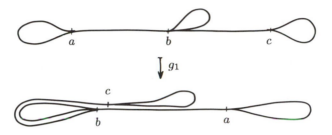

FIG. 42. Expanding map of a branched manifold. The map $g = g_1^2$ extends to a map $f : \mathbf{R}^2 \mapsto \mathbf{R}^2$ which, by perturbation, gives a diffeomorphism $\tilde{f}$ isotopic to the identity and with a strange attractor. (Using g_1 would give an orientation-reversing diffeomorphism.)

Since the action of $\tilde{f}$ on Λ is conjugate to the action of $g^\dagger$ on $S^\dagger$, it is described by symbolic dynamics as shown above.

D.11. Attractors of Topological Dimension 1 for Maps

In the last section, we based the construction of the solenoid on the doubling map of the circle. In a similar manner, one obtains an Axiom A attractor from any smooth expanding map g of a compact manifold S. This idea of Smale was pushed even further by Williams, who used expanding maps of branched manifolds instead of ordinary manifolds.

For simplicity, we discuss the 1-dimensional case. A 1-dimensional C^1 compact *branched manifold* S is obtained by gluing together a finite number of 1-*cells* (intervals of $\mathbf{R}$) at a finite number of branch points; the various branches make angles of 0 or π at branch points, so that the tangent bundle is well defined (see Fig. 41 for an example). C^1 functions, and a continuous Riemann metric, are defined on S as for an ordinary manifold.

An *expanding map* g of the compact connected 1-dimensional branched manifold S to itself is assumed to be smooth and to have the following properties:

(a) g expands in the sense of (D.3).

(b) each branch point has a neighborhood U that is flattened to a line interval by a suitable g^n, $n > 0$ (for instance, in the case of Fig. 41, U may be of the form $(AB) \cup (AB') \cup (AB'')$, and $g^n U$ some interval (CD).

(c) S is nonwandering.

Again we may embed S in a compact manifold M (we assume $\dim M \geq 4$), we may extend g to a map $f : M \mapsto M$ for which S is a pre-Axiom A basic set and then perturb f to a diffeomorphism $\tilde{f}$ of M so that S goes over into an Axiom A attractor Λ. The reader is asked to convince himself or herself that this construction can be made C^∞ instead of C^1. The unstable manifolds $\mathcal{V}_x^+ \subset \Lambda$ have dimension 1, and the sets $\mathcal{V}_x^- \cap \Lambda$ are Cantor sets, of topological dimension zero.[31] Therefore Λ has topological dimension 1. Conversely, Williams has shown[32] that if Λ^* is an Axiom A attractor of topological dimension 1 for the map f^*, then $f^*|\Lambda^*$ is topologically conjugate to $g^\dagger$ acting on the hyperbolic cover $S^\dagger$ of a branched manifold S, for an expanding map g.

Williams' method yields in particular a strange Axiom A attractor corresponding to a diffeomorphism $\tilde{f}$ *isotopic to the identity* on any 2-dimensional manifold M. ($\tilde{f}$ is isotopic to the identity if there is a continuous family $(\tilde{f}_\mu)_{0 \leqslant \mu \leqslant 1}$ of diffeomorphisms $\tilde{f}_\mu : M \mapsto M$ such that $\tilde{f}_0$ is the identity and $\tilde{f}_1 = \tilde{f}$). The existence of such attractors is not obvious and was first established by Plykin [1]; Figure 42 shows an example (cf. Newhouse, Ruelle, and Takens [1]).

D.12. The Stability Problem and the No-Cycle Condition

If Λ is an Axiom A basic set, a C^1 small perturbation of the dynamical system will lead to a new invariant set $\tilde{\Lambda}$ close to Λ; there is, in fact, a map $\Lambda \mapsto \tilde{\Lambda}$ that sends the orbits of the old system onto those of the new system. This is a *local stability property* which follows from

---

[31] See Appendix A.2 for the definition of the topological dimension.

[32] The proof of Williams [1] has to be supplemented by the following remark: By a small perturbation, f^* may be assumed C^2, and then the stable manifolds form a C^1 foliation (Section D.8). See also Williams [2] for the classification of 1-dimensional attractors, and Williams [3] for an extension to branched manifolds of higher dimension. See also Kollmer [1].

hyperbolicity (see Sections 15.5, 15.7).

Notice that a *local* Axiom A basic set Λ may be contained in a strictly larger Axiom A basic set Λ^*, for instance, if Λ^* is a horseshoe and Λ a periodic orbit. This, however, does not occur if Λ is an attractor [let V be an open set intersecting Λ and such that $\cup_{t\geq0} f^t V$ is contained in a fundamental neighborhood U of Λ, $\cup_{t\geq0} f^t V$ is dense in any basic set Λ^* containing Λ, and therefore $\Lambda^* = \Lambda$]. We are thus led to studying more generally the stability of a compact attracting set Λ (see Section 8.1); in particular if M is compact, we may take $\Lambda = M$.

We say that a compact attracting set Λ, with fundamental neighborhood U, satisfies Axiom A if the set Ω of nonwandering points in U is hyperbolic and if periodic points are dense in Ω. (Obviously $\Omega \subset \Lambda$.) The theory developed up to now applies here also. If Ω_i is a basic set, we may define

$$W_{\Omega_i}^- = \{x \in U : f^t x \to \Omega_i \text{ for } t \to +\infty\},$$
$$W_{\Omega_i}^+ = \{x_0 : (x_s)_{s\leq0} \text{ is an orbit contained in } U,$$
$$\text{and } x_s \to \Omega_i \text{ for } s \to -\infty\}.$$

It is not hard to see that *the $W_{\Omega_i}^-$ form a partition of U*, whereas *the $W_{\Omega_i}^+$ form a covering of* Λ. We write

$$\Omega_i \twoheadrightarrow \Omega_j \quad \text{if } W_{\Omega_i}^+ \cap W_{\Omega_i}^- \neq \emptyset.$$

The *No-Cycle* condition is that one cannot find distinct $\Omega_{i_1},\ldots,\Omega_{i_p}$, with $p > 1$, such that

$$\Omega_{i_1} \twoheadrightarrow \Omega_{i_2} \twoheadrightarrow \cdots \twoheadrightarrow \Omega_{i_p} \twoheadrightarrow \Omega_{i_1}.$$

We shall see (Theorem D.14) that Axiom A and the No-Cycle condition imply a *global stability property* for the nonwandering set. This is based on a geometric fact—the existence of filtrations—which we study first.

D.13. Proposition (Filtrations). *Let Λ be a compact attracting set with fundamental neighborhood U, satisfying Axiom A and the No-Cycle condition, and assume that the Ω_i, $i = 1,\ldots,N$ are ordered so that $\Omega_i \twoheadrightarrow \Omega_j$ implies $i > j$.*

(a) *The set $\Lambda_k = \bigcup_{i\leq k} W_{\Omega_i}^+$ is compact, and the set $W_k^- = \bigcup_{i\leq k} W_{\Omega_i}^-$ is an open neighborhood of Λ_k. We have the invariance properties $f^t \Lambda_k = \Lambda_k$, and for sufficiently large T, $f^T W_k^- \subset W_k^-$.*

(b) *The Λ_k are attracting sets, and they have open fundamental neighborhoods U_k such that $U_1 \subset \cdots \subset U_N$. We have $\Lambda_N = \Lambda$, so that we may replace U by U_N. Write $U_0 = \emptyset$, then*

$$\Omega_k = \{x_0 : (x_s)_{-\infty}^{\infty} \text{ is an orbit contained in } U_k \backslash U_{k-1}\}.$$

A sequence $U_0 \subset U_1 \subset \cdots \subset U_N$ with the properties of (b) is called a *filtration*[33] associated with the dynamical system (f^t).

To prove (a), we first note that

$$\Lambda_k = \bigcup_{i \leq k} \mathcal{W}_{\Omega_i}^+ \subset \bigcup_{i \leq k} \mathcal{W}_{\Omega_i}^- = \mathcal{W}_k^-$$

as a consequence of the definition of $\succ\!\!\!\!-$ and the ordering of the Ω_i. The invariance properties follow from $f^t \mathcal{W}_{\Omega_i}^+ = \mathcal{W}_{\Omega_i}^+$, $f^T \mathcal{W}_{\Omega_i}^- \subset \mathcal{W}_{\Omega_i}^-$. It remains to prove that Λ_k is closed and $\mathcal{W}_k^-$ is open.

Let x be in the closure of $\mathcal{W}_{\Omega_i}^+$. There are then orbits $(y_s^\varepsilon)_{-\infty}^0$ contained in $\mathcal{W}_{\Omega_i}^+$ such that $y_0^\varepsilon \to x$ when $\varepsilon \to 0$. In the limit $\varepsilon \to 0$, the orbit (y_s^ε) may come close to the basic sets $\Omega_i = \Omega_{j_0}, \Omega_{j_1}, \ldots, \Omega_{j_\ell} = \Omega_j$ (ordered by decreasing index). Using the compactness of Λ, one constructs orbits $(x_s^{(1)})_{-\infty}^{\infty}, \ldots, (x_s^{(\ell)})_{-\infty}^{\infty}$ and $(x_s^{(\ell+1)})_{-\infty}^0$ going from Ω_{j_0} to $\Omega_{j_1}, \ldots$, from $\Omega_{j_{\ell-1}}$ to Ω_{j_ℓ}, and from Ω_{j_ℓ} to x. Therefore, $x \in \mathcal{W}_{\Omega_j}^+$, $\Omega_j \succ\!\!\!\!- \Omega_{j_{\ell-1}} \succ\!\!\!\!- \cdots \succ\!\!\!\!- \Omega_i$ so that the closure of $\mathcal{W}_{\Omega_i}^+$ is contained in $\Lambda_i = \bigcup_{j \leq i} \mathcal{W}_{\Omega_j}^+$. This implies that Λ_k is closed. Similarly, one can show that the closure of $\mathcal{W}_{\Omega_i}^-$ is contained in $\bigcup_{j \geq 1} \mathcal{W}_{\Omega_j}^-$. Therefore, $\bigcup_{i > k} \mathcal{W}_{\Omega_i}^-$ is closed in U, so that $\mathcal{W}_k^-$ is open.

Note that if $x \in \Lambda$, we may write $x = x_0$, where $(x_s)_{-\infty}^0$ is an orbit in Λ. There is thus a subsequence $(x_{s(n)})$ tending to a nonwandering point y when $s(n) \to -\infty$. Let $y \in \Omega_i$, then $x_s \to \Omega_i$ when $s \to -\infty$, and therefore $x \in \mathcal{W}_{\Omega_i}^+$. We have thus shown that $\Lambda \subset \Lambda_N$, and hence $\Lambda = \Lambda_N$.

To complete the proof of (b), we shall construct the U_k by induction on increasing k. Suppose thus that the sets $U_0 \subset \cdots \subset U_{k-1}$ have the desired properties, and define

$$W = \{x \in \mathcal{W}_{\Omega_k}^+ : d(x, \Omega_k) < \varepsilon\}$$

---

[33] Perhaps one can trace the concept of filtration back to Morse in his study of gradient flows. For more general dynamical systems, see Smale [4]. Filtrations have proven to be a powerful tool; see, in particular, Conley [1].

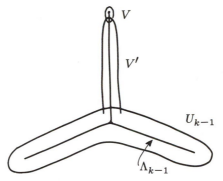

FIG. 43. Construction of a fundamental neighborhood in the proof of Proposition D.13.

for some (small) $\varepsilon > 0$. Since the set $\Lambda_k \backslash U_{k-1} = \mathcal{W}^+_{\Omega_k} \backslash U_{k-1}$ is compact, we may take T such that, if $t \geqslant T$,

$$ f^t W \supset \Lambda_k \backslash U_{k-1} $$

and

$$ f^t (\Lambda_k \backslash W) \subset U_{k-1}. $$

We may now choose $\delta, \delta' > 0$ such that if V' is the δ'-neighborhood of $\Lambda_k \backslash W$, then $f^t V' \subset U_{k-1}$, and if V is the δ-neighborhood of W, then $f^t V \subset V \cup V'$ for $t \geqslant T$ (see Fig. 43). [Here we use the fact that Ω_k is hyperbolic, and ε small: f compresses in a direction transversal to the unstable manifolds; we may assume that once an orbit $f^t x$ leaves the region of transversal compression, it never comes back near W].

Writing $U_k = U_{k-1} \cup V \cup V'$, we see that $f^t U_k \subset U_k$ for $t \geqslant T$, and in fact $f^t U_k$ is contained in an arbitrarily small neighborhood of Λ_k for k large enough, so that Λ_k is an attracting set.

By our construction, an orbit $(x_s)_{-\infty}^{\infty}$ contained in $U_k \backslash U_{k-1}$ is contained in V and therefore in Ω_k. This concludes the proof of the proposition.

D.14. Theorem (Ω-stability).[34] *Let Λ be a compact attracting set with fundamental neighborhood U for the C^1 map f or semiflow (f^t).*

---

[34]See Smale [3], [4] for diffeomorphisms, and Pugh and Shub [2] for flows on a compact manifold. A converse to this theorem is described in Section D.17.

(a) *It is equivalent to assume that the Axiom A and No-Cycle condition hold or that the chain recurrent set is hyperbolic. Under these conditions, the chain recurrent set is equal to the nonwandering set Ω.*

(b) *Let $\tilde{f}$ or $(\tilde{f}^t)$ be a C^1 small perturbation of the dynamical system f or (f^t). The perturbed system has a compact attracting set $\tilde{\Lambda}$ and nonwandering set $\tilde{\Omega}$, and there is a surjective continuous map $h : \Omega \mapsto \tilde{\Omega}$ sending orbits to orbits.*

First, assume that the chain-recurrent set R is hyperbolic, then the periodic points are dense in R in view of Problem 5 of Part 2. Thus $R = \Omega$, and Axiom A holds. Suppose that $\Omega' = \Omega_{i_1} \twoheadrightarrow \Omega_{i_2} \twoheadrightarrow \cdots \twoheadrightarrow \Omega_{i_p} = \Omega''$. If there existed $x \in \mathcal{W}_{\Omega'}^- \cap \mathcal{W}_{\Omega''}^+$ (existence of a cycle), then x would be chain recurrent, in contradiction to the fact that x cannot belong to any Ω_i (they are invariant and disjoint). We have thus shown that if R is hyperbolic, then Axiom A and the No-Cycle condition are satisfied. The converse implication follows readily from Proposition D.13.

To prove (b), we shall use the filtration $U_0 \subset U_1 \subset \cdots \subset U_N$ of Proposition D.13. By the stability of hyperbolic sets (Sections 15.5, 15.7), the perturbed system has prehyperbolic sets $\tilde{\Omega}_k$. If the perturbed system is sufficiently close to the old one, the orbits $(\tilde{x}_s)_{-\infty}^\infty$ that are entirely contained in $U_k \backslash U_{k-1}$ are in fact in $\tilde{\Omega}_k$.

Let $\tilde{\Lambda} = \bigcap_{t \geqslant 0} \tilde{f}^t U$. If $(\tilde{x}_s)_{-\infty}^\infty$ is an orbit in $\tilde{\Lambda}$ then, for some k, it will remain in $U_k \backslash U_{k-1}$ when $s \to -\infty$ and therefore be contained in the unstable manifold of $\tilde{\Omega}_k$. Let $\tilde{\Lambda}_k$ be the union of the unstable manifolds of $\tilde{\Omega}_1, \ldots, \tilde{\Omega}_k$. We show that $\tilde{\Lambda}_k$ is a compact attracting set by induction on increasing k. For each $\delta > 0$, the unstable manifold of $\tilde{\Omega}_k$ is the union of a compact set and a part contained in a δ-neighborhood of $\tilde{\Lambda}_{k-1}$; therefore, $\tilde{\Lambda}_k$ is compact. The proof that $\tilde{\Lambda}_k$ is an attracting set is like the corresponding proof for Λ_k in Proposition D.13. Finally, the nonwandering set is $\tilde{\Omega} = \tilde{\Omega}_1 \cup \cdots \cup \tilde{\Omega}_N$, concluding the proof of the theorem.

D.15. Strong Transversality

Let M be a compact manifold and (f^t) an Axiom A dynamical system generated by a diffeomorphism f of M, or a flow. If x is in the nonwandering set Ω, its global stable and unstable manifolds are

$$W_x^- = \{y \in M : \lim_{t \to \infty} d(f^t x, f^t y) = 0\},$$
$$W_x^+ = \{y \in M : \lim_{t \to -\infty} d(f^t x, f^t y) = 0\}.$$

For each basic set Ω_i, we can also define

$$W_{\Omega_i}^\pm = \{y \in M : \lim_{t \to \mp\infty} d(f^t y, \Omega_i) = 0\},$$

and we have

$$W_{\Omega_i}^\pm = \bigcup_{x \in \Omega_i} W_x^\pm$$

as a consequence of the existence of fundamental neighborhoods (Sections 15.5, 15.7).

The definition of the nonwandering set Ω implies that for every $y \in M$, $f^t y \to \Omega$ when $t \to \infty$. In fact, $f^t y \to \Omega_i$ for one of the basic sets Ω_i. [By invariance and compactness of the basic sets, $f^t y$ cannot jump from a neighborhood of Ω_i to another basic set and remain close to Ω all the time.] Therefore, the $W_{\Omega_i}^-$ form a partition of M, and it follows that the W_x^- also form a partition of M (two manifolds W_x^-, $W_{x'}^-$ are either disjoint or identical). Similar results hold for the global unstable manifolds $W_{\Omega_i}^+$, W_x^+.

We say that the dynamical system (f^t) satisfies *strong transversality* if, for each $y \in M$, the manifolds W_x^+, $W_{x'}^-$ through y are such that

$$T_y M = T_y W_x^+ + T_y W_{x'}^- (+X),$$

where X is added in the flow case if y is not a fixed point and X is the direction of the flow (i.e., $X = \mathbf{R}\frac{d}{dt} f^t y$). In other words, it is assumed that, at the point y, W_x^+ and $W_{x'}^-$ are transversal in the case of a diffeomorphism, and that W_x^{0+} and $W_{x'}^-$ are transversal in the case of a flow.

Strong transversality implies the No-Cycle condition, i.e., there cannot be a sequence

(D.4) $$\Omega_{i_1} \twoheadrightarrow \Omega_{i_2} \twoheadrightarrow \cdots \twoheadrightarrow \Omega_{i_p} \twoheadrightarrow \Omega_{i_1},$$

where

$$\Omega_i \twoheadrightarrow \Omega_j \quad \text{means} \quad W_{\Omega_i}^+ \cap W_{\Omega_j}^- \neq \emptyset$$

(see Section D.12). Let us sketch the proof in the diffeomorphism case. By replacing f by a suitable iterate f^N, we may assume that f is topologically mixing on the Ω_i. Then, if x is a periodic point in Ω_i, $W_x^{\pm}$ are dense in Ω_i [this follows from Theorem D.4]. Therefore, taking periodic points $x_i \in \Omega_{i_1}, \ldots, x_{i_p} \in \Omega_{i_p}$, we see (by using (D.4) and strong transversality) that

$$W_{x_i}^- \cap W_{x_{i+1}}^+ \ni y_i \quad \text{for } i = 1, \ldots, p-1,$$
$$W_{x_p}^- \cap W_{i_1}^+ \ni y_p,$$

where $y_1, \ldots, y_p$ are points of transversal intersection. Therefore $y_1, \ldots, y_p$ are nonwandering. [This is the *cloud lemma* we discussed for the case $p = 2$ in the proof of Theorem D.2.]

D.16. Structural Stability

In this section, M will again be a *compact* manifold; we investigate structurally stable diffeomorphisms and flows on M (see Section 8.6).

Remember that the C^r diffeomorphisms form an open subset $\text{Diff}^r(M)$ of the space $C^r(M, M)$ of C^r maps of M to itself (Section 2.7). The diffeomorphism f is C^r-structurally stable if, for g sufficiently close to f in $\text{Diff}^r(M)$, there is a homeomorphism $h : M \mapsto M$ such that

$$h \circ f = g \circ h.$$

We have defined the space $\mathcal{F}^r(M)$ of C^r flows on M in Section 2.7. In particular, $\mathcal{F}^r(M)$ contains the flows associated with C^r vector fields on M. The flow (f^t) is C^r-structurally stable if, for (g^t) sufficiently close to (f^t) in $\mathcal{F}^r(M)$, there is a homeomorphism $h : M \mapsto M$ that maps f-orbits to g-orbits, preserving the order of points on orbits. Note that C^1 structural stability of a C^r diffeomorphism or flow implies C^r structural stability for $r \geqslant 1$.

If an Axiom A diffeomorphism $f \in \mathcal{D}^1(M)$ or flow $(f^t) \in \mathcal{F}^1(M)$ satisfies strong transversality, then it is C^1-structurally stable.

The proof of this theorem is not easy; it is due to Robbin [1] and Robinson [2], [1]. For a converse, see the next section.

D.17. Characterization of Structurally Stable and Ω-stable Systems

It is believed that structurally stable and Ω-stable diffeomorphisms and flows on a compact manifold are precisely those encountered in Sections

D.16 and D.14 above. More precisely, one has the following conjectures (Palis, Smale).

C^r **stability conjecture.**   The diffeomorphism f is C^r structurally stable if and only if f satisfies Axiom A and strong transversality; similarly for flows.

$C^r \Omega$-**stability conjecture.**   The diffeomorphism f is C^r Ω-stable if and only if f satisfies Axiom A and the No-Cycle condition; similarly for flows.

Mañé [2] has proven the C^1 stability conjecture for diffeomorphisms. This is a difficult theorem; the methods do not extend to $r > 1$, and the case of flows has not been treated. Based on Mañé's results, Palis [6] has proven the C^1 Ω-stability conjecture for diffeomorphisms. Again, this does not extend to $r > 1$, and the case of flows has not been treated.

D.18. Anosov Diffeomorphisms and Flows

If f is a diffeomorphism of the compact manifold M, and if M is hyperbolic with respect to f, then f is said to be an *Anosov diffeomorphism*. Similarly, if (f^t) is a flow on the compact manifold M, and if M is hyperbolic with respect to (f^t), then (f^t) is said to be an *Anosov flow*. Such dynamical systems satisfy Axiom A and strong transversality and are therefore structurally stable.

It is not known if, for an Anosov diffeomorphism f of M, the non-wandering set Ω necessarily coincides with M. There is, however, an Anosov flow on a 3-manifold such that the chain-recurrent set is strictly smaller than M (see Francks and Williams [1]).

Let A be an $m \times m$ matrix with integer coefficients, and $\det A = \pm 1$; then A determines a diffeomorphism $\widetilde{A}$ of $\mathbf{R}^n / \mathbf{Z}^n$ (*toral automorphism*). If the spectrum of A is disjoint from $\{z : |z| = 1\}$, then $\widetilde{A}$ is an Anosov diffeomorphism.

Let N be a Riemann manifold and M the set of unit tangent vectors to N. The frictionless motion of a point mass with unit velocity on N is a flow on M called *geodesic flow*. If the sectional curvature of N is everywhere negative, the geodesic flow is Anosov.

The above examples explain the early interest in Anosov diffeomorphisms and flows (see Anosov [1]). Anosov systems have served as prototypes for later studies of more general hyperbolic systems (see Smale [3]).

D.19. Note

Usually, the discussion of Axiom A dynamical systems is restricted to diffeomorphisms and flows on compact manifolds (basic references are Smale [3], and Bowen [6]). Here we have tried to relax these conditions. To pursue the study of Axiom A systems beyond the geometric treatment given in the present Appendix, it is necessary to use *Markov partitions* and *symbolic dynamics*. (These were introduced by Sinai [1], [2], [3], and Bowen [1], [4]; for a general discussion of these topics, see Bowen [5], and also Ruelle [2].) Closely resembling an Axiom A system is the *geometric Lorenz attractor* (for which we refer to Guckenheimer [2], Guckenheimer and Williams [1], and Williams [4]). The study of nonuniformly hyperbolic dynamical systems in general requires ergodic theory, and is rather different in spirit from the geometric considerations of the present monograph (see Oseledec [1], Pesin [1], [2], [3], Ruelle [1], Katok [1], Ledrappier and Young [1], etc.).

References

[1] Abraham, R., and Marsden, J. E. *Foundations of Mechanics.* 2nd ed. Benjamin/Cummings, Reading, Mass., 1978.

[1] Abraham, R., and Robbin, J. *Transversal Mappings and Flows.* Benjamin, New York, 1967.

[1] Abraham, R., and Shaw, C. *Dynamics—The Geometry of Behavior. I. Periodic Behavior. II. Chaotic Behavior. III. Global Behavior.* Aerial Press, Santa Cruz, Calif., 1982, 1983, 1984.

[1] Anosov, D. V. "Geodesic flows on a compact Riemann manifold of negative curvature," *Trudy Mat. Inst. Steklov* **90** (1967). English translation, *Proc. Steklov Math. Inst.* **90** (1967).

[1] Arnold, V. I. "Small denominators. I. Mappings of the circumference onto itself," *Izv. Akad. Nauk SSSR, Ser. Mat.* **25** (1961), 21–86. English translation, *Amer. Math. Soc. Transl.* **46** (1965), 213–284.

[1] Belitskii, G. R. "Functional equations and conjugacy of local diffeomorphisms of a finite smoothness class," *Functional Anal. Appl.* **7** (1973), 268–277.

[1] Berger, M. S. *Nonlinearity and Functional Analysis.* Academic Press, New York, 1977.

[1] Bonic, R., and Frampton, J. "Smooth functions on Banach manifolds," *J. Math. Mech.* **15** (1966), 877–898.

[1] Bourbaki, N. *Théorie des Ensembles. (Fascicule de Résultats).* Hermann, Paris, 1958.

[2] Bourbaki, N. *Topologie Générale. (Fascicule de Résultats).* Hermann, Paris, 1953.

[3] Bourbaki, N. *Espaces Vectoriels Topologiques. (Fascicule de Résultats).* Hermann, Paris, 1955.

[4] Bourbaki, N. *Intégration.* Chapitres 1, 2, 3, 4. Hermann, Paris, 1965.

[5] Bourbaki, N. *Variétés Différentielles et Analytiques. (Fascicule de Résultats).* Hermann, Paris, 1967 (I), 1971 (II).

[1] Bowen, R. "Markov partitions for axiom A diffeomorphisms," *Amer. J. Math.* **92** (1970), 725–747.

[2] Bowen, R. "Periodic points and measures for axiom A diffeomorphisms," *Trans. Amer. Math. Soc.* **154** (1971), 377–397.

[3] Bowen, R. "Periodic orbits for hyperbolic flows," *Amer. J. Math.* **94** (1972), 1–30.

[4] Bowen, R. "Symbolic dynamics for hyperbolic flows," *Amer. J. Math.* **95** (1973), 429–459.

[5] Bowen, R. *Equilibrium States and the Ergodic Theory of Anosov Diffeomorphisms.* Lecture Notes in Math. **470**, Springer-Verlag, Berlin, 1975.

[6] Bowen, R. *On Axiom A Diffeomorphisms.* CBMS Regional Conference Series **35**, Amer. Math. Soc., Providence, R.I., 1978.

[1] Broer, H. W., Huitema, G. B., and Takens, F. "Unfolding of quasiperiodic tori," *Groningen ZW-8705,* Preprint 1987.

[1] Campanino, M., and Epstein, H. "On the existence of Feigenbaum's fixed point," *Commun. Math. Phys.* **79** (1981), 261–302.

[1] Campanino, M., Epstein, H., and Ruelle, D. "On Feigenbaum's functional equation $g \circ g(\lambda x) + \lambda g(x) = 0$," *Topology* **21** (1982), 125–129.

[1] Carr, J. *Applications of Centre Manifold Theory.* Applied Math. Sci. **35**, Springer-Verlag, New York, 1981.

[1] Chenciner, A. "Bifurcations de points fixes elliptiques. I. Courbes invariantes," *Publ. Math. IHES* **61** (1985), 67–127. II. "Orbites périodiques et ensembles de Cantor invariants," *Invent. Math.* **80** (1985), 81–106. III. "Orbites périodiques de "petites" périodes et

élimination résonnante des couples de courbes invariantes," *Publ. Math. IHES* **66** (1988), 5–91.

[1] Choquet, G. "Itération de difféomorphismes; attracteurs globaux et ponctuels," *C. R. Acad. Sc. Paris* **305** Sér. I (1987), 617–621.

[1] Choquet, G., and Meyer, P.-A. "Existence et unicité des représentations intégrales dans les convexes compacts quelconques," *Ann. Inst. Fourier* **13** (1963), 139–154.

[1] Chossat, P., and Golubitsky, M. "Iterates of maps with symmetry," *SIAM J. Math. Anal.* **19** (1988).

[1] Collet, P., and Eckmann, J.-P. *Iterated Maps on the Interval as Dynamical Systems.* Progress in Physics, Vol. 1. Birkhäuser, Boston, 1980.

[1] Collet, P., Eckmann, J.-P., and Koch, H. "Period doubling bifurcations for families of maps on $\mathbf{R}^n$," *J. Statist. Phys.* **25** (1981), 1–14.

[1] Conley, Ch. *Isolated Invariant Sets and the Morse Index.* CBMS Regional Conference Series **38**, Amer. Math. Soc., Providence, R.I., 1978.

[1] Curry, J. H. "On the Hénon transformation," *Commun. Math. Phys.* **68** (1979), 129–140.

[1] Cvitanović, P. *Universality in Chaos,* Adam Hilger, Bristol, 1984.

[1] Danker, A. "On Smale's axiom A dynamical systems," *Annals of Math.* **107** (1978), 517–553.

[1] De Rham, G. *Variétés Différentiables.* Hermann, Paris, 1960.

[1] Denker, M., Grillenberger, C., and Sigmund, K. *Ergodic Theory on Compact Spaces.* Lecture Notes in Math. **527**, Springer-Verlag, Berlin, 1976.

[1] Duhem, P. *La Théorie Physique. Son Object et sa Structure.* Chevalier et Rivière, Paris, 1906.

[1] Eckmann, J.-P. "Roads to turbulence in dissipative dynamical systems," *Rev. Mod. Phys.* **53** (1981), 643–654.

[1] Eckmann, J.-P., and Ruelle, D. "Ergodic theory of chaos and strange attractors," *Rev. Mod. Phys.* **57** (1985), 617–656.

[1] Epstein, H. "New proofs of the existence of the Feigenbaum functions," *Commun. Math. Phys.* **106** (1986), 395–426.

[1] Erber, T., Guralnik, S. A., and Latal, H. G. "A general phenomenology of hysteresis," *Ann. Phys.* **69** (1972), 161–192.

[1] Feigenbaum, M. J. "Quantitative universality for a class of nonlinear transformations," *J. Statist. Phys.* **19** (1978), 25–52.

[2] Feigenbaum, M. J. "The universal metric properties of nonlinear transformations," *J. Statist. Phys.* **21** (1979), 669–706.

[1] Feigenbaum, M., Kadanoff, L., and Shenker S. "Quasiperiodicity in dissipative systems, a renormalization group analysis," *Physica* **5D** (1982), 370–386.

[1] Feit, S. D. "Characteristic exponents and strange attractors," *Commun. Math. Phys.* **61** (1978), 249–260.

[1] Fenichel, N. "Persistence and smoothness of invariant manifolds for flows," *Indiana Univ. Math. J.* **21** (1971), 193–226.

[1] Franks, J., and Williams, R. F. "Anomalous Anosov flows," in: *Global Theory of Dynamical Systems.* Lecture Notes in Math. **819**, Springer-Verlag, Berlin, 1980, 158–174.

[1] Golubitsky, M., and Stewart, I. "Hopf bifurcation in the presence of symmetry," *Arch. Rational Mech. Anal.* **87** (1985), 107–165.

[1] Grebogi, C., McDonald, S., Ott, E., and Yorke, J. "Final state sensitivity: an obstruction to predictability," *Phys. Lett.* **99**A (1983), 415–418.

[1] Grebogi, C., Ott, E., and Yorke, J. "Chaotic attractors in crisis," *Phys. Rev. Let.* **48** (1982), 1507–1510.

[1] Grobman, D. M. "Topological classification of the neighborhood of a singular point in n-dimensional space," *Mat. Sbornik (N. S.)* **56** (98) N° 1 (1962), 77–94.

[1] Guckenheimer, J. "Endomorphisms of the Riemann sphere," in: *Global Analysis.* Proc. Symp. Pure Math. **14** (1970), 95–124.

[2] Guckenheimer, J. "A strange, strange attractor," in: J. E. Marsden and M. McCracken [1], 368–381.

[1] Guckenheimer, J., and Holmes, Ph. *Nonlinear Oscillations, Dynamical Systems, and Bifurcations of Vector Fields.* Applied Math. Sci. **42**, Springer-Verlag, New York, 1983.

[1] Guckenheimer, J., and Williams, R. F. "Structural stability of Lorenz attractors," *Publ. Math. IHES* **50** (1979), 59–72.

[1] Hadamard, J. "Les surfaces à courbures opposées et leurs lignes géodésiques," *J. Math. Pures et Appl.* **4** (1898), 27–73.

[1] Halmos, H. *Measure Theory.* D. van Nostrand, Princeton, N.J., 1950.

[1] Hao Bai-Lin. *Chaos.* World Scientific, Singapore, 1984.

[1] Hartman, P. "A lemma in the theory of structural stability of differential equations," *Proc. Amer. Math. Soc.* **11** (1960), 610–620.

[1] Hénon, M. "A two dimensional mapping with a strange attractor," *Commun. Math. Phys.* **50** (1976), 69–77.

[1] Herman, M. "Sur la conjugaison différentiable des difféomorphismes du cercle à des rotations," *Publ. Math. IHES* **49** (1979), 5–233.

[2] Herman, M. "Simple proofs of local conjugacy theorems for diffeomorphisms of the circle with almost every rotation number," *Bol. Soc. Bras. Mat.* **16** (1985), 45–83.

[1] Hirsch, M. W. "Expanding maps and transformation groups," in: *Global Analysis*. Proc. Symp. Pure Math. **14** (1970), 125–131.

[2] Hirsch, M. W. *Differential Topology*. Springer-Verlag, New York, 1976.

[1] Hirsch, M., Palis, J., Pugh, C., and Shub, M. "Neighborhoods of hyperbolic sets," *Invent. Math.* **9** (1970), 121–134.

[1] Hirsch, M., and Pugh, C. "Stable Manifolds and hyperbolic sets," in: *Proc. Symp. in Pure Math.* **14**, Amer. Math. Soc., Providence, R.I., 1970, 133–164.

[1] Hirsch, M., Pugh, C., and Shub, M. *Invariant Manifolds*. Lecture Notes in Math. **583**, Springer-Verlag, Berlin, 1977.

[1] Hopf, E. "Abzweigung einer periodischen Lösung von einer stationären Lösung eines Differentialsystems," *Ber. Math.-Phys. Kl. Sächs. Akad. Wiss. Leipzig* **94** (1942), 1–22.

[2] Hopf, E. "A mathematical example displaying the features of turbulence," *Comm. Pure Appl. Math.* **1** (1948), 303–322.

[1] Iooss, G. *Bifurcation of Maps and Applications*. Lecture Notes, Mathematical Studies. North-Holland, Amsterdam, 1979.

[1] Irwin, M. C. "On the stable manifold theorem," *Bull. London Math. Soc.* **2** (1970), 196–198.

[1] Jacobs, K. *Lecture Notes On Ergodic Theory*. Aarhus Universitet, Aarhus, 1963.

[1] Katok, A. "Lyapunov exponents, entropy, and periodic orbits for diffeomorphisms," *Publ. Math. IHES* **51** (1980), 137–174.

[1] Kelley, A. "The stable, center-stable, center, center-unstable, and unstable manifolds," Appendix C in: R. Abraham and J. Robbin [1], 134–154.

[1] Kelley, J. L. *General Topology*. Springer-Verlag, New York, 1955.

[1] Khanin, K. M., and Sinai, Ya. G. "A new proof of M. Herman's theorem," *Commun. Math. Phys.* **112** (1987), 89–101.

[1] Kollmer, H. "On hyperbolic attractors of codimension one," in: *Geometry and Topology III*, Lecture Notes in Math., **597** Springer-Verlag, Berlin, 1977, 330–334.

[1] Kupka, I. "Contribution à la théorie des champs génériques," *Contrib. Diff. Equations* **2** (1963), 457–484.

[1] Kurata, M. "Hyperbolic nonwandering sets without dense periodic points," *Nagoya Math. J.* **74** (1979), 77–86.

[1] Landau, L. D. "On the problem of turbulence," *Dokl. Akad. Nauk SSSR* **44**, 8 (1944), 339–342.

[1] Lanford, O. E. "Bifurcation of periodic solutions into invariant tori," in: *Nonlinear Problems in the Physical Sciences and Biology.* Lecture Notes in Math. **322**, Springer-Verlag, Berlin, 1973, 159–192.

[2] Lanford, O. E. "A computer-assisted proof of the Feigenbaum conjectures," *Bull. Amer. Math. Soc.* **6** (1982), 427–434.

[3] Lanford, O. E. "Introduction to the mathematical theory of dynamical systems," in: *Chaotic Behavior of Deterministic Systems.* (Les Houches 1981) North-Holland, Amsterdam, 1983, 3–51.

[1] Lang, S. *Differential Manifolds.* Addison-Wesley, Reading, Mass., 1972.

[1] Ledrappier, F., and Young, L.-S. "The metric entropy of diffeomorphisms. Part I. Characterization of measures satisfying Pesin's formula. Part II. Relations between entropy, exponents and dimension," *Ann. of Math.* **122** (1985) 509–539, 540–574.

[1] Lorenz, E. N. "Deterministic nonperiodic flow," *J. Atmos. Sci.* **20** (1963), 130–141.

[1] Mañé, R. "Lyapounov exponents and stable manifolds for compact transformations," in: *Geometric Dynamics.* Lecture Notes in Math. **1007**, Springer-Verlag, Berlin, 1983, 522–577.

[2] Mañé, R. "A proof of the C^1 stability conjecture," *Publ. Math. IHES* **66** (1988), 161–210.

[1] Marsden, J. E., and McCracken, M. *The Hopf Bifurcation and Its Applications.* Applied Math. Sci. **19**, Springer-Verlag, New York, 1976.

[1] Mestel, B. D. "Computer assisted proof of universality for cubic critical maps of the circle with golden mean rotation number," Thesis, Warwick University, 1985.

[1] Milnor, J. "On manifolds homeomorphic to the 7-sphere," *Ann. of Math.* **64** (1956), 399–405.

[1] Moise, E. E. *Geometric Topology in Dimensions 2 and 3.* Springer-Verlag, New York, 1977.

[1] Moser, J. *Stable and Random Motions in Dynamical Systems.* Ann. Math. Studies **77**. Princeton Univ. Press, Princeton, N.J., 1973.

[1] Neĭmark, Ju. I. "On some cases of dependence of periodic solutions on parameters," *Dokl. Akad. Nauk SSSR* **129**, 4 (1959), 736–739.

[1] Nelson, E. *Topics in Dynamics I: Flows.* Princeton University Press, Princeton, N.J., 1969.

[1] Newhouse, S. "Diffeomorphisms with infinitely many sinks," *Topology* **13** (1974), 9–18.

[2] Newhouse, S. "The abundance of wild hyperbolic sets and non-smooth stable sets for diffeomorphisms," *Publ. Math. IHES* **50** (1979), 102–151.

[3] Newhouse, S. "Lectures on dynamical systems," in: *CIME Lectures, Bressanone, Italy, June 1978.* Birkhäuser, Boston, 1980, 1–114.

[1] Newhouse, S., and Palis, J. "Bifurcations of Morse–Smale dynamical systems," in: *Dynamical Systems* (ed. M. Peixoto). Academic Press, New York, 1973, 303–366.

[2] Newhouse, S., and Palis, J. "Cycles and bifurcation theory," *Asterisque* **31** (1976), 43–141.

[1] Newhouse, S., Palis, J., and Takens, F. "Bifurcation and stability of families of diffeomorphisms," *Publ. Math. IHES* **57** (1983), 5–72.

[1] Newhouse, S., Ruelle, D., and Takens, F. "Occurrence of strange axiom A attractors near quasiperiodic flows on T^m, $m \geqslant 3$," *Commun. Math. Phys.* **64** (1978), 35–40.

[1] Nihon Sugakkai. *Encyclopedic Dictionary of Mathematics* (1st Engl. ed. 2 vols.), 1987 (2nd English ed. 4 vols.), MIT Press, Cambridge, Mass., 1977 .

[1] Nitecki, Z. *Differentiable Dynamics. An Introduction to the Orbit Structure of Diffeomorphisms.* MIT Press, Cambridge, Mass., 1971.

[1] Oseledec, V. I. "A multiplicative ergodic theorem. Lyapunov characteristic numbers for dynamical systems," *Trudy Moskov. Mat. Obšč.* **19** (1968), 179–210. English translation, *Trans. Moscow Math. Soc.* **19** (1968), 197–221.

[1] Ostlund, S., Rand, D. A., Sethna, J., and Siggia, E. D. "Universal properties of the transition from quasi-periodicity to chaos in dissipative systems," *Physica* **8**D (1983), 303–342.

[1] Palis, J. "On the local structure of hyperbolic points in Banach spaces," *Anais Acad. Brasil. Ciências* **40** (1968), 263–266.

[2] Palis, J. "On Morse–Smale dynamical systems," *Topology* **8** (1969), 385–405.

[3] Palis, J. "A note on Ω-stability," in: *Global Analysis*. Proc. Symp. Pure Math. **14**, Amer. Math. Soc., Providence, R.I., 1970, 221–222.

[4] Palis, J. "Ω-Explosions," *Proc. Amer. Math. Soc.* **27** (1971), 85–90.

[5] Palis, J. "A note on the inclination lemma (λ-lemma) and Feigenbaum's rate of approach" in: *Geometric Dynamics* Lecture Notes in Math. **1007**, Springer-Verlag, Berlin, 1983, 630–636.

[6] Palis, J. "On the C^1 Ω-stability conjecture," Publ. Math. IHES **66** (1988), 211–215.

[1] Palis, J., and de Melo, W. *Geometric Theory of Dynamical Systems*. Springer-Verlag, New York, 1982.

[1] Palis, J., and Takens, F. "Topological equivalence of normally hyperbolic dynamical systems," *Topology* **16** (1977), 335–345.

[2] Palis, J., and Takens, F. "Cycles and measure of bifurcation sets for two-dimensional diffeomorphisms," *Invent. Math.* **82** (1985), 397–422.

[3] Palis, J., and Takens, F. "Hyperbolicity and the creation of homoclinic orbits," *Ann. of Math.* **125** (1987), 337–374.

[1] Pesin, Ya. B. "Lyapunov characteristic exponents and ergodic properties of smooth dynamical systems with an invariant measure," *Dokl. Akad. Nauk SSSR* **226**, 4 (1976), 774–777. English translation, *Soviet Math. Dokl.* **17**, 1 (1976), 196–199.

[2] Pesin, Ya. B. "Invariant manifold families which correspond to nonvanishing characteristic exponents," *Izv. Akad. Nauk SSSR Ser. Mat.* **40**, 6 (1976), 1332–1379. English translation, *Math. USSR Izv.* **10**, 6 (1976), 1261–1305.

[3] Pesin, Ya. B. "Lyapunov characteristic exponents and smooth ergodic theory," *Uspehi Mat. Nauk* **32**, 4 (196), (1977), 55–112. English translation, *Russian Math. Surveys* **32**, 4 (1977), 55–114.

[1] Plante, J. "Anosov flows," *Amer. J. Math.* **94**, 729–754 (1972).

[1] Plykin, R. V. "Sources and currents of A-diffeomorphisms of surfaces," *Math. Sbornik* **94**, 2(6), (1974), 243–264.

[1] Poincaré, H. "Sur une classe étendue de transcendantes uniformes," *CRAS Paris* **103** (1886), 862–864.

[2] Poincaré, H. "Sur une classe nouvelle de transcendantes uniformes," *J. de Math.* 4e *Sér.* **6** (1890), 313–365.

[3] Poincaré, H. "Les nouvelles méthodes de la mécanique céleste," Gauthier–Villars, Paris, 1892 (I), 1893 (II), 1899 (III).

[4] Poincaré, H. *Science et Méthode* Ernest Flammarion, Paris, 1908.

[5] Poincaré, H. *Oeuvres,* Tome IV. Gauthier-Villars, Paris, 1950.

[1] Pomeau, Y., and Manneville, P. "Intermittent transition to turbulence in dissipative dynamical systems," *Commun. Math. Phys.* **74** (1980), 189–197.

[1] Pugh, C. "An improved closing lemma and a general density theorem," *Amer. J. Math.* **89** (1967), 1010–1021.

[2] Pugh, C. "On a theorem of P. Hartman," *Amer. J. Math.* **91** (1969), 363–367.

[1] Pugh, C., and Robinson, C. "The C^1 closing lemma, including Hamiltonians," *Ergod. Th. and Dynam. Syst.* **3** (1983), 261–313.

[1] Pugh, C., and Shub, M. "Linearization of normally hyperbolic diffeomorphisms and flows," *Invent. Math.* **10** (1970), 187–198.

[2] Pugh, C., and Shub, M. "The Ω-stability theorem for flows," *Invent. Math.* **11** (1970), 150–158.

[1] Rand, D. "Dynamics and symmetry: predictions for modulated waves in rotating waves," *Arch. Rational Mech. Anal.* **79** (1982), 1–38.

[1] Reed, M., and Simon, B. *Methods of Modern Mathematical Physics.* Academic Press, New York, 1972 (I), 1975 (II), 1979 (III), 1978 (IV).

[1] Renz, P. "Equivalent flows on smooth Banach manifolds," *Indiana Univ. Math. J.* **20** (1971), 695–698.

[1] Robbin, J. W. "A structural stability theorem," *Ann. Math.* **94** (1971), 447–493.

[1] Robinson, C. "Structural stability of C^1 flows," in: *Dynamical Systems—Warwick 1974,* Lecture Notes in Math **468**, Springer-Verlag, Berlin, 1975, 262–277.

[2] Robinson, C. "Structural stability of C^1 diffeomorphisms," *J. Diff. Equ.* **22** (1976), 28–73.

[3] Robinson, C. "Bifurcation to infinitely many sinks," *Commun. Math. Phys.* **90** (1983), 433–459.

[1] Ruelle, D. "Bifurcations in the presence of a symmetry group," *Arch. Rational Mech. Anal.* **51** (1973), 136–152.

[2] Ruelle, D. *Thermodynamic Formalism.* Encyclopedia of Math. and its Appl., vol. 5. Addison-Wesley, Reading, Mass., 1978.

[3] Ruelle, D. "Ergodic theory of differentiable dynamical systems," *Publ. Math. IHES* **50** (1979), 27–58.

[4] Ruelle, D. "Small random perturbations of dynamical systems and the definition of attractors," *Commun. Math. Phys.* **82** (1981), 137–151.

[5] Ruelle, D. "Characteristic exponents and invariant manifolds in Hilbert space," *Ann. Math.* **115** (1982), 243–290.

[6] Ruelle, D. "Repellers for real analytic maps," *Ergod. Th. and Dynam. Syst.* **2** (1982), 99–107.

[1] Ruelle, D., and Takens, F. "On the nature of turbulence," *Commun. Math. Phys.* **20** (1971), 167–192. Note concerning our paper "On the nature of turbulence," *Commun. Math. Phys.* **23** (1971), 343–344.

[1] Sacker, R. J. "On invariant surfaces and bifurcation of periodic solutions of ordinary differential equations," Thesis, NYU, 1964, unpublished.

[1] Sattinger, D. H. *Group Theoretic Methods in Bifurcation Theory.* Lecture Notes in Math. **762**, Springer-Verlag, Berlin, 1979.

[1] Shub, M. "Endomorphisms of compact differentiable manifolds," *Amer. J. Math.* **91** (1969), 175–199.

[2] Shub, M. "Stabilité globale des systèmes dynamiques," *Astérisque* **56** (1978), 1–211. English translation, *Global Stability of Dynamical Systems.* Springer-Verlag, New York, 1987.

[1] Siegel, C. L. "Über die Normalform analytischer Differentialgleichungen in der Nähe einer Gleichgewichtslösung," *Nachr. Akad. Wiss. Göttingen. Math.-physikal. Kl.* **5** (1952), 21–30.

[1] Šil'nikov, L. P. "A contribution to the problem of the structure of an extended neighborhood of a rough equilibrium state of saddle-focus type," *Mat. Sbornik* **81** (**123**) (1970), 92–103. English translation, *Math. USSR Sbornik* **10** (1970), 91–102.

[1] Sinai, Ia. G. "Markov partitions and C-diffeomorphisms," *Funkts. Analiz i ego Pril.* **2**, 1 (1968), 64–89. English translation, *Functional Anal. Appl.* **2** (1968), 61–82.

[2] Sinai, Ia. G. "Construction of Markov partitions," *Funkts. Analiz i ego Pril.* **2**, 3 (1968), 70–80. English translation, *Functional Anal. Appl.* **2** (1968), 245–253.

[3] Sinai Ia. G. "Gibbs measures in ergodic theory," *Uspeki Mat. Nauk* **27**, 4 (1972), 21–64. English translation, *Russian Math. Surveys* **27**, 4 (1972), 21–69.

[1] Smale. S. "Stable manifolds for differential equations and diffeomorphisms," *Ann. Scuola Norm. Sup. Pisa* **17** (1963), 97–116.

[2] Smale, S. "Diffeomorphisms with many periodic points," in: *Differential and Combinatorial Topology*. Princeton U. Press, Princeton, N.J., 1965, 63–80.

[3] Smale, S. "Differentiable dynamical systems," *Bull. Amer. Math. Soc.* **73** (1967), 747–817.

[4] Smale, S. "The Ω-stability theorem," in: *Global Analysis*. Proc. Symp. Pure Math. **14**, Amer. Math. Soc., Providence, R.I., 1970, 289–298.

[1] Sternberg, S. "Local contractions and a theorem of Poincaré," *Amer. J. Math.* **79** (1957), 809–824.

[2] Sternberg, S. "On the structure of local homeomorphisms of Euclidean n-space II," *Amer. J. Math.* **80** (1958), 623–631.

[1] Takens, F. "Partially hyperbolic fixed points," *Topology* **10** (1971), 133–147.

[2] Takens, F. "Singularities of vector fields," *Publ. Math. IHES* **43** (1974), 47–100.

[3] Takens, F. "A note on the differentiability of centre manifolds." in: *Dynamical Systems and Partial Differential Equations: Proceedings of the VII ELAM* Ed. Equinoccio, Caracas, 1986, 101–104.

[1] Thom, R. *Stabilité Structurelle et Morphogènise*. Benjamin, Reading, Mass., 1972.

[1] Tresser, C. "Un théorème de Šil'nikov en $C^{1,1}$," *CRAS Paris* **296** Ser. I (1983), 545–548.

[2] Tresser, C. "About some theorems by L. P. Šil'nikov," *Ann. Inst. Henri Poincaré* **40** (1984), 441–461.

[1] Walters, P. *Ergodic Theory. Introductory Lectures*. Lecture Notes in Mathematics **458**, Springer-Verlag, Berlin, 1975.

[1] Williams, R. F. "One-dimensional nonwandering sets," *Topology* **6** (1967), 473–487.

[2] Williams, R. F. "Classification of one dimensional attractors," in: *Global Analysis*. Proc. Symp. Pure Math. **14**, Amer. Math. Soc., Providence, R.I., 1970, 341–361.

[3] Williams, R. F. "Expanding attractors," *Publ. Math. IHES* **43** (1974), 169–203.

[4] Williams, R. F. "The structure of Lorenz attractors," *Publ. Math. IHES* **50** (1979), 73–99.

[1] Yoccoz, J.-C. "Conjugaison différentiable des difféomorphismes du cercle dont le nombre de rotation vérifie une condition diophantienne," *Ann. Scient. E.N.S 4ᵉ Sér.* **17** (1984), 333–359.

Index